*Stolen
From the office of
Neel C. Row*

PROGRESS IN PUMPS

PROGRESS IN PUMPS

Edited by

Jay Matley

and

The Staff of Chemical Engineering

McGraw-Hill Information Services Co., New York, N.Y.

Copyright © 1989 by Chemical Engineering McGraw-Hill Information Services Co.
1221 Avenue of the Americas, New York, New York 10020

All rights reserved. No parts of this work may be reproduced or utilized in any form or by any means, electronic or mechanical, including photocopying, microfilm and recording, or by any information storage and retrieval system without permission in writing from the publisher.

Printed in the United States of America.

Library of Congress Cataloging-in-Publication Data

Progress in pumps / edited by Jay Matley and the staff of Chemical
    engineering
        p.   cm.
    Includes index.
    1. Pumping machinery. I. Matley, Jay. II. Chemical engineering.
TJ900.P73 1989
621.6'9--dc20   89-12136
                                                                     CIP

ISBN 0-07-607009-3

CONTENTS

Introduction		vii
Part I	**Advances in pump technology**	1
	Progress in pumps	3
	New pump models aim at satisfying CPI needs	16
Part II	**Centrifugal pumps**	21
	Downtime prompts upgrading of centrifual pumps	23
	Standard pumps are *not* obsolete!	30
	Head-vs.-capacity characteristics of centrifugal pumps	32
	Multistage centrifugal pumps	35
	Time takes its toll on centrifugal pumps	38
	Designing centrifugal pump systems	41
	Pump bypasses now more important	45
	Centrifugal pumps and system hydraulics	50
	Unusual problems with centrifugal pumps	73
	Startup of centrifugal pumps in flashing or cryogenic liquid service	76
	Choosing plastic pumps	78
Part III	**Positive-displacement pumps**	81
	Selecting positive-displacement pumps	83
	Predicting flowrates from positive-displacement rotary pumps	98
	Alternative to Gaede's formula for vacuum pumpdown time	102
Part IV	**Mechanical seals**	105
	A users' guide to mechanical seals	107
	Call for higher quality is heeded by seal makers	114
	When to select a sealless pump	118
	Seals for abrasive slurries	122
	Power consumption of double mechanical seals	127
	Troubleshooting mechanical seals	130

Part V Pump drives — 141

New turn for CPI motors — 143
Specifying electric motors — 147
Check pump performance from motor data — 159
Making the proper choice of adjustable-speed drives — 161

Index — 175

INTRODUCTION

The past half-dozen years have been particularly eventual ones with regard to pump technology. Competitive pressures on both chemical processors and equipment manufacturers have brought about significant changes in the design and construction of pumps, in efforts to boost efficiency.

In the past, engineers could specify some overcapacity in pumps, for safety margins, because the penalties in capital and operating costs were not objectionable. This is no longer the case. In past years, engineers were not hard pressed keeping abreast of pump technology. This, too, is no longer the case. Developments in pump technology are occurring at a more rapid pace than ever before.

Centrifugal and positive-displacement pumps are being redesigned and their components are being constructed of new materials. New types of pumps are being introduced. Many of these changes are described in the opening article, from which this book gets its title. Demands for greater reliability have prompted calls in some quarters for changes in the ANSI (American National Standards Institute) and API (American Petroleum Institute) standards, with few counterclaims that revisions are not necessary. As of July 1989, the revision of API610 has been completed and that of ANSI B73.1 is approaching completion.

Pump accessories—particularly seals and drives—have received their share of attention. Part of the reasons for seeking changes in the ANSI and API standards are attributed to relaxing, or altering, dimensional standards so that pumps can be equipped with better seals, which are available. Greater concerns about safety have generated interest in leak-free pumps, boosting demand for seal-less pumps. Efforts to reduce operating costs have focused renewed attention upon variable-speed drives for pumps.

Thus, the design and construction of pumps for the chemical process industries has undergone unprecedented turmoil, which is showing no signs of subsiding soon. This book is indispensable for the engineer who needs to catch up with the changes that have occurred and to gauge the directions of future changes.

Part I
Advances in pump technology

Progress in pumps
New pump models aim at satisfying CPI needs

PROGRESS IN PUMPS

Significant advances in pump technology have taken place recently, and more can be expected in the immediate years ahead. Let us review some developments of recent years and their application to the chemical process industries (CPI).

Basic principles remain

Pumping principles — the way pumps perform and the equations for calculating or estimating pump performance — have not changed significantly, except for certain areas where additional research has added new understanding. Table I presents a brief review of some of the more-often-used pump formulas, both in customary U.S. units and in conventional metric and SI (Système International d'Unités) metric units. U.S. engineers are still reluctant to switch to SI; the change must eventually be made, and the sooner the better. Pump manufacturers in Europe and the Far East have been using metric units for quite some time.

There is still some difference of opinion about which metric units to use for pressure and flow. SI standardizes on kilopascals (kPa) as the unit of pressure; some European and Asian engineers prefer pressure in bar or kg/cm^2. The most common unit of flow is cubic meters per hour (m^3/h). But for most pumps in the CPI, the hourly rate is too clumsy, and we find pumps being selected inaccurately as a result.

SI literature suggests that capacities be stated in m^3/h, but for calculations, volume/second is preferred. The suggested unit is cubic decimeters per second (dm^3/s), essentially liters per second, which gets pump capacity into a more understandable range. Table I offers formulas employing both of these unit variations. The international pump industry would be wise to agree upon one group of standard units for pumps.

There is increasing awareness by both sellers and users of the need to select more-efficient pumps in process services. Forward-looking engineers recognize the waste in oversizing pumps by too great a margin.

It has been traditional in specifying pumps to add a safety margin, or overcapacity, to the net required flow. Perhaps this idea originates in the process designer's mind, as that person may not be totally sure of the required flow. But it may also be true that organizations continue to use such a safety factor from force of habit.

Richard F. Neerken
The Ralph M. Parsons Co.

The past few years have seen many changes in the kinds of centrifugals made and bought, changes in the ways of specifying pumps, new pump standards, new seals, and even some new pump types. Here is a rundown of what has happened.

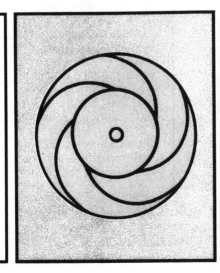

Originally published September 14, 1987

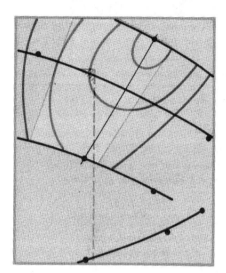

Example 1 illustrates both the traditional and the "energy saving" way to select a pump for process duty, so that efficiency is highest at the normal operating point. Pump A is slightly larger than Pump B. Pump A is selected with the rated point to the left of the best efficiency point (B.E.P.). Pump B is selected past the B.E.P., resulting in a power saving when pumping at the normal process rate. With acceptance of this concept, users may alter the statement in many pump specifications that "...pump rated-capacity point shall be at or to the left of the best efficiency point for the selected pump," replacing it with, "... pump rated-capacity point shall be no more than 5 or 10% beyond the B.E.P." Example 1 shows why it is so often cost effective to have the "rated point" slightly to the *right* of the best efficiency point (B.E.P.)

Trends in pump usage

After many years of noting the requirements and desires of major pump-users, I see an increasing preference for:
- More vertical inline centrifugal pumps (to save space).
- Very large sizes of pumps (rather than multiple smaller ones), where systems require them and such designs are available.

- Choosing or specifying types of pumps other than centrifugals for certain low-flow or very-high-head applications (to get improved efficiency).
- Computer-aided methods for pump selection, preparation of pump quotations, data sheets and curves, requisition-orders writing, control of parts inventories, and bills of material for shop fabrication and assembly of each machine.

Vertical inline centrifugal pumps have been popular since their introduction more than 25 years ago. Some users are reluctant to change, however, and may place limits on where verticals may be used. For example, many will consider vertical inline types only to a maximum of, say, 100 hp. The vertical type saves space, which should mean lower first cost of a new facility. Designs have been improved with regard to shaft and bearing support by new designs that incorporate a bearing housing for the pump impeller, rather than use the motor bearings to support it. This concept has caused some of the early troubles with shaft deflection and misalignment

to disappear. Example 2 shows calculations for shaft deflections in three styles of vertical inline pumps, and indicates why new designs are preferred.

Higher-than-60-Hz speeds (i.e., >3,600 rpm) are now accepted by most users, resulting in a better match of hydraulic performance than was previously possible in low-flow, high-head applications, with 3,600-rpm limitations.

Some types have limits, however — being currently available in flow to about 400 gpm, or in power to about 350 bhp. Direct-connected, 3,600-rpm vertical inline pumps have been built to sizes up to about 1,800 gpm at 600-ft head, or for considerably higher flows at lower heads, when running at 1,750, 1,150 or 880 rpm.

Use of computerized methods for pump-related work is increasing. Some pump makers are now using computer methods for printing proposals or proposal datasheets. Nearly every major manufacturer uses computers in pump design, material control, inventory, bills of materials, and numerically controlled manufacturing processes. Spare-parts orders are often handled by computers. Many pump companies are now using computer-aided design and drafting methods. Several computerized pump pricing methods are in use [1,2], and, in addition, at least one contracting firm has a pump selection program designed to help with preliminary selections. Type of pump, duty conditions, metallurgy, seal requirements, driver, and test details are input data. The computer "looks" at the curves of each manufacturer of the specified type of pump. With permission of the participating pump companies, curve data has been stored so that the program searches for one or more suitable pumps only within the region where the contractor's engineer has decided it should look. Then it interpolates between lines of various impeller diameters and pump power requirements to match the input duty conditions. In medium to large size pumps, the engineer may limit the region to be considered to a narrow band around the best efficiency area (about -10% and +5% of B.E.P.). The computer then prints out which pumps, if any, will meet the duty requirements; calculates performance, price and weight (in accordance with current price-list data in the computer memory); and then prints all of this information.

After selection of the most suitable pump by the engineer, the computer prints the pump specification sheet, or data

Table I — Here are the formulas commonly used for determining power, specific speed, net positive suction head, impeller speed, and other key calculations

Customary U.S. units	Metric or SI units
For determination of pump power requirements	
$[BHP] = \dfrac{\text{Flow, gpm} \times \text{head, ft} \times \text{sp.gr.}}{3{,}960 \times \text{efficiency (decimal)}}$	$KW = \dfrac{\text{dm}^3/\text{s} \times \text{head, m} \times \text{density}}{102{,}000 \times \text{efficiency (decimal)}}$
	$KW = \dfrac{\text{m}^3/\text{h} \times \text{head, m} \times \text{density}}{3.6 \times 102{,}000 \times \text{efficiency (decimal)}}$
$[BHP] = \dfrac{\text{Flow, gpm} \times \text{differential pressure, psi}}{1{,}715 \times \text{efficiency, (decimal)}}$	$KW = \dfrac{\text{dm}^3/\text{s} \times \text{differential pressure, kPa}}{1{,}000 \times \text{efficiency (decimal)}}$
	$KW = \dfrac{\text{m}^3/\text{h} \times \text{differential pressure, bar}}{3.6 \times 10 \times \text{efficiency (decimal)}}$
For conversion of pressure to head	
$\text{Head, ft} = \dfrac{\text{pressure, psi} \times 2.31}{\text{specific gravity}}$	$\text{Head, m} = \dfrac{\text{pressure, kPa} \times 102}{\text{density, kg/m}^3}$
	or $\text{Head, m} = \dfrac{\text{pressure, bar} \times 10{,}200}{\text{density, kg/m}^3}$
(Atmospheric pressure at sea level = 14.696 psia = 34 ft of water)	(Atmospheric pressure at sea level = 101.3 kPa = 1.013 bar = 10.34 m of water)
(34/14.696 = 2.31)	$\left(\dfrac{10.34 \times 1{,}000}{101.3} = 102\right)$
For determination of specific speed	
$N_s = \dfrac{\text{rpm} \sqrt{(\text{flow, gpm})}}{\text{Head, ft}^{3/4}}$	$N_{ss} = \dfrac{\text{rpm} \sqrt{(\text{flow, dm}^3/\text{s})}}{\text{Head, m}^{3/4}}$
or, for suction specific speed:	
$N_{ss} = \dfrac{\text{rpm} \sqrt{(\text{flow, gpm})}}{\text{NPSHR, ft}^{3/4}}$	$N_s = \dfrac{\text{rpm} \sqrt{(\text{flow, dm}^3/\text{s})}}{\text{NPSHR, m}^{3/4}}$

Note: For double suction pumps, use one-half of total flow when calculating suction specific speed. Always use flow at pump's best efficiency point, with maximum dia. impeller.

sheet, and a price estimate sheet. Just another step away is printing of the preliminary outline showing dimensions required for layout or even for inclusion in 3-dimensional CAD (computer-aided design) of complete systems. We expect a great increase in such computer-aided methods.

We foresee the appearance soon of systems by which users or contractors will be able to "talk" to pump manufacturers' computers, obtain needed information directly transmitted by computer, and possibly eliminate or streamline the costly, time-consuming bidding process. Through such methods, more time can be spent in study of system hydraulics for each pump circuit, and the making of thorough, cost-effective choices for a given duty requirement. Fig. 1 illustrates typical input and output sheets from such a program, and also shows an actual pump curve, highlighting the area where the program was instructed to look.

Low-flow cavitation

The phenomenon of cavitation in centrifugal pumps has called forth an enormous amount of research and discussion

---

Table I — Continued

Customary U.S. units	Metric or SI units
For determination of NPSH available:	
NPSHA = pressure at source − pipe friction in suction line to pump − vapor pressure of liquid at pumping temperature + static head of liquid above pump inlet	
Note: All units must be the same; either convert pressures into heads (i.e., NPSHA in height units), or convert static head into pressure (NPSHA in pressure units).	
Relation of speed to capacity and head	**Relation of impeller diameter to capacity and head**
$\dfrac{N_1}{N_2} = \dfrac{Q_1}{Q_2} = \sqrt{\dfrac{H_1}{H_2}}$	$\dfrac{D_1}{D_2} = \dfrac{Q_1}{Q_2} = \sqrt{\dfrac{H_1}{H_2}}$
Note: Points 1 and 2 must be taken through constant efficiency lines on pump curve.	
For determination of impeller tip speed	
$U_2,\ \text{ft/s} = \dfrac{\text{Dia.(in.)} \times \pi \times \text{rpm}}{(60)(12)}$ or $= \dfrac{\text{Dia.(in.)} \times \text{rpm}}{229}$	$U_2,\ \text{m/s} = \dfrac{\text{Dia.(mm)} \times \pi \times \text{rpm}}{(1,000)(60)}$ or $= \dfrac{\text{Dia.(mm)} \times \text{rpm}}{19,098}$
For determination of shaft deflection (overhung pumps)	
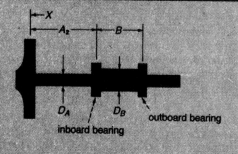	$Y = \dfrac{W}{3E}\left(\dfrac{A_2^3}{I_A} + \dfrac{A_2^3 B}{I_B}\right) - \left(\dfrac{W}{3E} + \dfrac{3A_2^2}{2I_A} + \dfrac{A_2 B}{I_B} - \dfrac{X^3}{2I_A}\right)$ E = shaft modulus of elasticity X = at any point along the shaft I = moment of inertia $(\pi d^4)/64$ Y = shaft deflection W = radial load
For determination of torque	
Torque, ft lb = $\dfrac{\text{BHP} \times 5{,}252}{\text{rpm}}$	Torque, Nm = $\dfrac{\text{KW} \times 9{,}549}{\text{rpm}}$

for more than fifty years. Early efforts to predict methods for ensuring stable, cavitation-free operation of pumps resulted in industrywide acceptance of 3% head decay, determined by a suppression test, as the criterion for cavitation limits of a pump.

Users are required to design pumping systems with realistic amounts of available NPSH (net positive suction head), and pass on the NPSH data to the pump manufacturers during the selection and bidding period. The manufacturers, in turn, are expected to furnish pumps that require an NPSH somewhat below the available amount, thus ensuring adequate pump life.

As pumps became larger and more complex, manufacturers and users found that the comparatively simple suppression-test methods did not always explain what would actually occur in pumping systems. Based on research and testing in the previous decade, the 1980s have seen industrywide technical discussions to share new ideas and test-methods that have helped designers predict system requirements to yield cavitation-free operation. Certain large users, confronted with some unusual pumping requirements for very large, high-energy pumps, began to investigate special laboratory testing of such pumps in an effort to achieve up to five years of cavitation-free operation.

The industry has seen the appearance of flow-visualization testing of prototype pumps or components, to help designers predict more accurately what kind of margin between NPSHA (NPSH available) and NPSHR (NPSH required) must be designed into the system. For large pumps, such five-year cavitation-free margins may be up to three or four times the traditional NPSHR values, as determined by the 3% head-decay suppression test.

Concurrent with this research and development, the phenomenon of low-flow cavitation was observed, especially in larger pumps, or in pumps with high suction-specific-speed (10,000 to 12,000, and above, in U.S. units — see Table I). Numerous papers, articles and book chapters have contained reports of such developments and testing, resulting in some interesting data to help the pump user design a troublefree system. At the risk of oversimplifying the subject, let us examine a real example of how a pump was incorrectly applied, and consequently exhibited low-flow cavitation.

It has become clear that in a real pump there is a margin between the NPSH required for cavitation-free operation (conveniently agreed to be equal to 40,000 h or 5 yr, with no significant impeller wear due to cavitation) and the NPSH required to make the pump cavitate, as determined in the traditional 3% method referred to before.

There appears to be an intermediate level, known as onset of cavitation, or incipient cavitation. See Ref. 6, p. 61, for a discussion of how and where this incipient cavitation begins. Fig. 2 illustrates how these several lines would appear for a real pump. It is clear that these levels are different for pumps of differing designs, and also that pump tests on water, in test laboratories, will not always indicate how the same pumps perform on actual fluids, such as hydrocarbon mixtures or chemical solutions, in plant processes. Some materials have been successfully pumped for years, indicating that the damage due to cavitation is not as great as if the same pump was handling cold water. Research has also confirmed that the convenient parameter of suction specific speed can help the user be aware of those pumps that may be potential troublemakers in a particular application.

Example 3 is taken from an older pump installation in potable water service in a petroleum refinery. The system is shown in Example 3a and the pump curve in 3b.

At first glance, this appears to be about as straightforward a system as one might wish to see; how could anything go wrong here? After operation began, however, the process flow required from this pump was decreased drastically

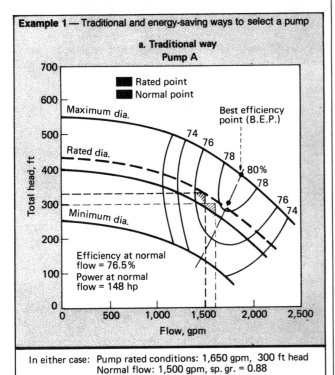
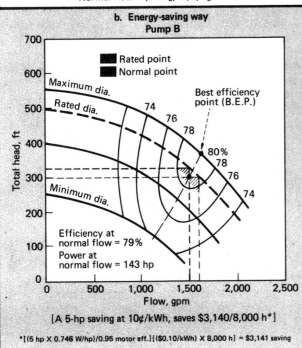

Example 1 — Traditional and energy-saving ways to select a pump

from the rated flow. (The rated flow had been based on startup needs and the user failed to recognize that once the system was full, the makeup water required would be much lower.) The pump operated for many months near the manufacturer's limit for minimum continuous stable flow.

The project was built before the API-610 Sixth Ed. had appeared, which contains, for the first time, a definition for "minimum continuous flow," viz.:

"Minimum continuous stable flow is the lowest flow rate at which the pump can operate without exceeding the noise and vibration limits imposed by this standard." (Para. 1.4.1).

Pump companies used to use the phrase "minimum flow" to mean a short or temporary flow during startup or shutdown. This user, however, meant "minimum continuous flow," somewhat in accord with the API definition.

During operation, the pump showed unusual noise. Finally, after four or five months of low-flow operation, a cast-iron impeller failed completely in one of the pumps. The pump had a design suction-specific-speed (at B.E.P.) of over 12,000, indicating that the actual shape of the pump NPSHR curve was more like Curve (2) than Curve (1) in Fig. 2, which had been determined by a suppression test. The problems were first corrected by installing a new impeller and a suitably sized bypass around the pump, to allow greater flow to pass through, at the cost of wasted power during the low-flow system requirements. Later, a smaller size pump was installed.

Fig. 3 shows a calculation example of this high N_{ss} pump impeller, where actual dimensions were measured at the site after low-flow cavitation had been observed. Using the charts from Refs. 3, 4 and 8, and having the measured dimensions called for by these charts, we can see that the predicted minimum flow on water was somewhere between 1,300 and 1,450 gpm. The manufacturer had originally quoted about 420 gpm minimum continuous flow!

Unfortunately, all this happened before the recent information was made available. But, today, this exemplifies the current dilemma for a new project: the detailed dimensions for proposed impellers are unavailable during the proposal stage unless very special requests are made to the bidders. On the other hand, the pump bidders have little or no knowledge of what the user needs for minimum *continuous* flow operation, nor does the bidder have detailed knowledge of the properties of the pumped liquid (beyond what the user may provide). The general comment that minimum flows based on water [3,8] can be reduced to 60% of the chart values for hydrocarbons, may not always be correct, depending on the composition of the pumped liquid. Laboratory tests for every pump are not practical, but this whole subject may call for more specialized pump testing when large, critical service, high suction-specific-speed pumps are to be used.

Other similar situations have been observed, all in pump installations

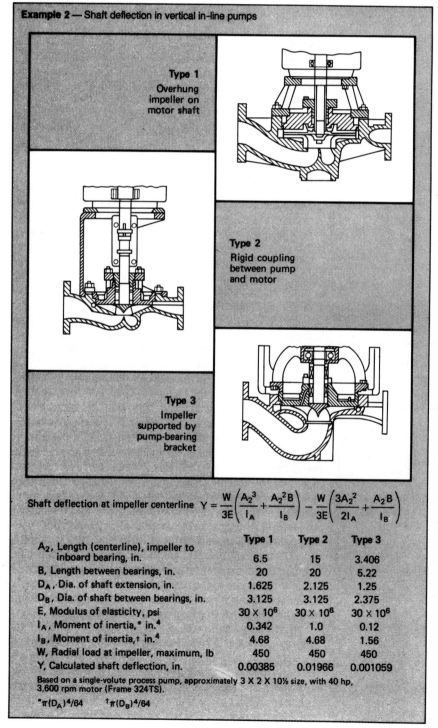

Example 2 — Shaft deflection in vertical in-line pumps

Type 1 — Overhung impeller on motor shaft

Type 2 — Rigid coupling between pump and motor

Type 3 — Impeller supported by pump-bearing bracket

Shaft deflection at impeller centerline $Y = \dfrac{W}{3E}\left(\dfrac{A_2^3}{I_A} + \dfrac{A_2^2 B}{I_B}\right) - \dfrac{W}{3E}\left(\dfrac{3A_2^2}{2I_A} + \dfrac{A_2 B}{I_B}\right)$

	Type 1	Type 2	Type 3
A_2, Length (centerline), impeller to inboard bearing, in.	6.5	15	3.406
B, Length between bearings, in.	20	20	5.22
D_A, Dia. of shaft extension, in.	1.625	2.125	1.25
D_B, Dia. of shaft between bearings, in.	3.125	3.125	2.375
E, Modulus of elasticity, psi	30×10^6	30×10^6	30×10^6
I_A, Moment of inertia,* in.⁴	0.342	1.0	0.12
I_B, Moment of inertia,† in.⁴	4.68	4.68	1.56
W, Radial load at impeller, maximum, lb	450	450	450
Y, Calculated shaft deflection, in.	0.00385	0.01966	0.001059

Based on a single-volute process pump, approximately 3 × 2 × 10½ size, with 40 hp, 3,600 rpm motor (Frame 324TS).

*$\pi(D_A)^4/64$ †$\pi(D_B)^4/64$

PROGRESS IN PUMPS

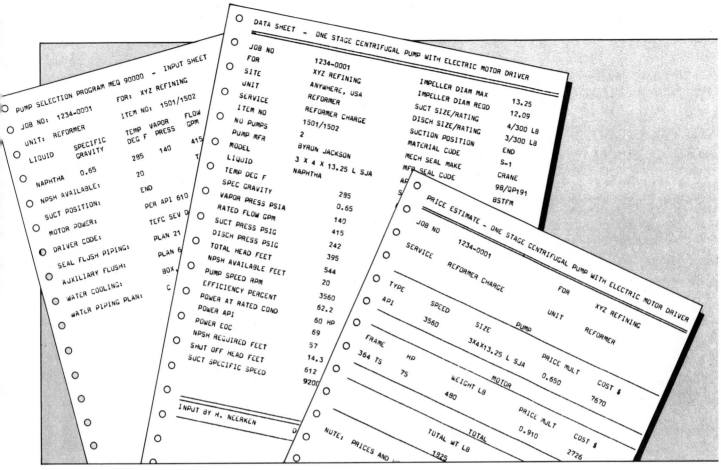

Figure 1 — A pump-selection program, directed to check within the shaded area of the pump curve, typically generates input and

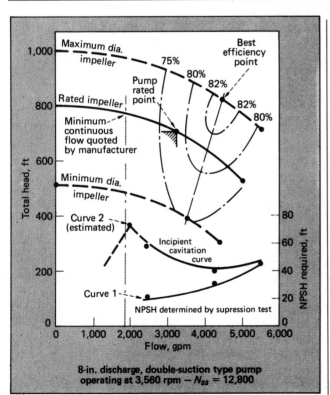

Figure 2 — Head needed to avoid cavitation, via the suppression test (for 3% head decay), and incipient cavitation

where $N_{ss} = 10,000$ or more. Table II shows a few of these. Pump manufacturers seem to be more aware of the difference between the 3% water-suppression-test requirements and the real pump performance, and are beginning to watch with greater care the suction specific speeds of pumps quoted for low-NPSHA systems or low minimum continuous-flow requirements.

This again shows an increasing need for more communication between pump seller and pump user, especially on larger, more-sophisticated systems. Users should bring the manufacturers' representatives into the system study and let them see and understand what will be expected of their pumps. Many annoying and costly problems can thus be avoided. Although a great deal of information on this subject of low-flow recirculation has appeared in the last ten years, the state-of-the-art knowledge still does not offer a complete troublefree simple correlation that is universally applicable. Each pump design requires special attention, especially designs that absorb higher energy.

API-610 Sixth Ed. also cites the need for greater care on "high energy pumps." Paragraph 2.1.10 states:

"High energy pumps (head greater than 650 feet per stage and more than 300 horsepower per stage) require special consideration to avoid blade passing frequency vibrations and low-frequency vibrations at reduced flow rates..."

Prior to this warning, many pump users neglected to pay special attention to large pumps that absorbed more energy per impeller, and that were possible candidates for trouble

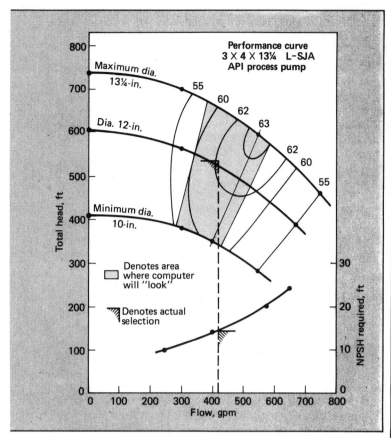

output sheets such as those shown

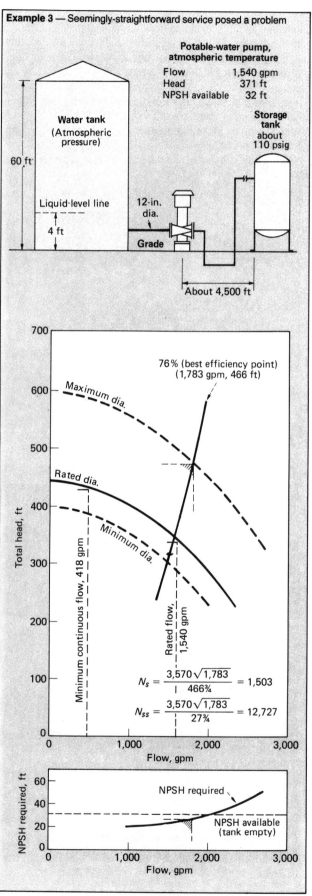

after installation. Such potential problems can be avoided, for example, by using pumps with more than one stage, to stay comfortably away from this high energy per stage input. For some applications, more-detailed manufacturer's shop testing of high-energy pumps will be required.

Use of pumps at other than 60-Hz speeds

More pumps are being applied at speeds above 3,600 rpm, the maximum speed for a 2-pole, 60-Hz motor. Larger pumps in pipeline and boiler-feed applications have been in operation for many years, with speeds in the 4,000 to 6,000-rpm range being fairly common. Some of these might be looked at as "high energy pumps" and have been successfully engineered and operated even before the API definition.

The integral-gear single-stage design appeared in the early 1960s, with speeds up to 20,000 rpm. The original concept was a vertical inline arrangement with built-in speed increaser. This type is limited to power requirements of about 250 to 400 hp. Horizontal shaft models are also available in one- and two-stage designs. A new variation has appeared in the 1980s, with horizontal shaft, plus a special, patented, low-flow-recirculation device in the pump suction. Steam- or gas-turbine drivers have been applied to medium and larger size pumps at speeds above 3,600 rpm, some direct connected, not requiring speed-increaser gears.

Motor drivers for such units may still require gears, but new adjustable-frequency motors are now widely available, and may eventually offer more-efficient pumping at higher speeds. Ref. 9 contains an excellent discussion of adjustable-speed drivers, including a.c. adjustable-frequency motors.

For pumps and drivers at speeds above 3,600 rpm, careful attention must be paid to types of bearings, seals, shaft support length, L/D (length/diameter) ratio, critical speed, and vibration response. Rolling element bearings (ball and roller types) are used in pumps up to about 4,000 to 5,000 rpm with reasonable assurance of long life. API-610 offers guidelines to determine whether hydrodynamic bearings are required. Paragraph 2.9.1.6 states: "Hydrodynamic radial and/or thrust bearings shall be required under any of the following conditions:

1. With barrel pumps.
2. When DN factors are 300,000 or greater. [The DN factor is the product of bearing size (bore) in millimeters and the rated speed in revolutions per minute.]
3. When standard antifriction bearings fail to meet an L-10 rating life of either 25,000 hours in continuous operation at rated conditions or 16,000 hours at maximum axial and radial loads and rated speed.
4. When the product of pump rated horsepower and rated speed (in revolutions per minute) is 2.7 million or greater."

Thus, in a small- to medium-size, multistage, split-case pump, for example, with a shaft dia. at the bearings of 2.5 in. (63.5 mm), the DN factor would equal or exceed 300,000 if the rotating shaft speed were greater than 4724 rpm. On larger multistage pumps, the result of this check nearly always calls for hydrodynamic bearings.

On small, vertical- or horizontal-shaft high-speed pumps, the DN factor might be below the 300,000 limit. (For example, a 1.25-in. shaft at 8,000 rpm equals a DN factor of 254,000 — but Item 3 or 4 of the API guide might be exceeded.) All available designs for the high-speed, partial emission-type pump employ sleeve bearings as standard practice. The pitot-tube pump (discussed later) is built with ball bearings, however, and in standard sizes is limited to about 5,400 rpm.

Velocity of the rotating seal faces must also be considered. For satisfactory mechanical-seal life, such values should be held to about 70 ft/s, or below, if conventional seals are used, up to 3,600 rpm pump speeds.

High-speed seals for pumps up to 8,000 rpm or higher (seal-face velocities to 150 to 200 ft/s) are available, and have been successfully applied to many high-speed charge pumps, pipeline pumps, and large boiler-feed pumps.

Shaft span between bearings, and shaft-span-to-bearing-diameter ratio (L/D) will make it possible to estimate the critical speed of the pump shaft. A typical comparison is shown in Table III. In this example, the high-speed 4-stage pump is a better choice than the conventional 3,600 rpm pump. The L/D ratios are shown, but the calculation of critical speed is complex in a multistage pump, and should not be attempted without a great deal of pump-design information not normally available to the user.

In short, the user is well advised to pay special attention to any pumps, large or small, that are to operate above 60-Hz speed. But with the present amount of experience with higher-speed pumps throughout the pump industry, users should not be reluctant to consider them, as they often represent the best engineering solution to the pumping problem (see Table III).

NPSH requirements for higher-speed pumps will necessarily be greater. Examination of the suction-specific-speed equation (Table I) will show why.

Several designs of high-speed pumps have inducer-type impellers, with a low-suction-specific-speed pump impeller

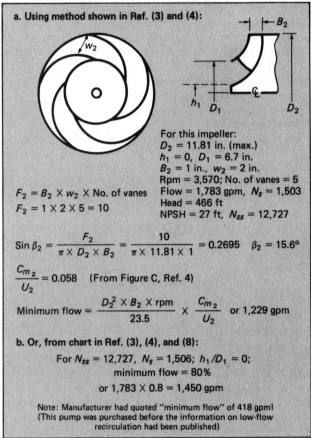

a. Using method shown in Ref. (3) and (4):

For this impeller:
D_2 = 11.81 in. (max.)
h_1 = 0, D_1 = 6.7 in.
B_2 = 1 in., w_2 = 2 in.
Rpm = 3,570; No. of vanes = 5
Flow = 1,783 gpm, N_s = 1,503
Head = 466 ft
NPSH = 27 ft, N_{ss} = 12,727

$F_2 = B_2 \times w_2 \times$ No. of vanes
$F_2 = 1 \times 2 \times 5 = 10$

$\sin \beta_2 = \dfrac{F_2}{\pi \times D_2 \times B_2} = \dfrac{10}{\pi \times 11.81 \times 1} = 0.2695 \quad \beta_2 = 15.6°$

$\dfrac{C_{m_2}}{U_2} = 0.058$ (From Figure C, Ref. 4)

Minimum flow = $\dfrac{D_2^2 \times B_2 \times \text{rpm}}{23.5} \times \dfrac{C_{m_2}}{U_2}$ or 1,229 gpm

b. Or, from chart in Ref. (3), (4), and (8):
For N_{ss} = 12,727, N_s = 1,506; h_1/D_1 = 0;
minimum flow = 80%
or 1,783 × 0.8 = 1,450 gpm

Note: Manufacturer had quoted "minimum flow" of 418 gpm! (This pump was purchased before the information on low-flow recirculation had been published)

Figure 3 — Predicted minimum flow can be dramatically different from manufacturer's quoted minimum flow

Table II — Low-flow cavitation can be a problem with some pumps having high suction specific-speeds

Liquid	Flow rated, gpm	Flow, minimum continuous	Total head, ft	N_{ss}	Trouble
Water	1,540	420	370	12,500	Noisy, excessive vibration, broken impeller
Hydrocarbon	4,200	1,500	210	15,000	No trouble
Hydrocarbon†	3,200	1,800	720	12,800	Noisy, excessive vibration, cavitation damage
Hydrocarbon†	19,000	2,200	410	12,000	Noisy at 5,000 gpm flow, no cavitation damage
Chemical solution†	1,300	550	620	11,200	Rotor shuttling (axial movement)
Hydrocarbon†	2,600	530	1,000	12,700	No trouble

*N_{ss} at best efficiency point, with maximum impeller. As quoted by pump manufacturer.
†Double suction type impeller.

(inducer) ahead of the main impeller, on the same shaft or as part of the same rotating assembly. Other, larger, designs will require separate booster pumps to provide enough NPSH for a high-speed main pump. Occasionally, plant layout will permit pump suction from a vessel located quite high above the pump; but, generally, to keep total costs as low as possible, column and vessel skirt heights must be lower, and use of inducer impellers or booster pumps may prove to be less costly than raising the NPSHA. A potential application for an inducer or booster pump must be reviewed

Table III — Shaft-span-to-bearing-diameter ratio (L/D) can differ markedly for high-speed and conventional pumps

Service — Hydrocracker charge	
Pumped liquid	Charge oil
Temperature, °F	300
Specific gravity	0.85
Pressure differential, psi	2,170
Head, ft	5,900
Flow, gpm	1,300
NPSH available	Ample

Type	High-speed pump Double case (barrel)	Conventional pump Double case (barrel)
Size, in.	4	4
No. stages	4	9
Speed, rpm	6,500	3,570
Efficiency, %	70	73
Power, kW	1,750	1,680
Shaft dia., D, in.	3.15	3.5
Shaft span, L, in.	52.2	95
L/D*	16.57	27.14

*Comparative factor for shaft deflection

with care, to be sure that alternative off-design conditions, minimum flow conditions, etc., will not cause trouble.

Finally, the future process system may be designed for even greater energy savings by eliminating the traditional control valve and its required pressure drop. This will be accomplished by use of the solid-state adjustable-frequency motor that can accept process-flow signals and respond directly by increasing or decreasing pump speed. Example 4 illustrates the possible energy saving in a typical CPI process installation that would show a reasonably fast payout of the added cost for the special drive system.

Pump sealing systems

Sealing of pump shafts to prevent leakage of product has likewise seen considerable advances in the 1980s. Increased pollution-control requirements in parts of the U.S. have mandated use of tandem or double mechanical-seals on certain pumps in CPI applications where vapor pressure at pumping temperature is high, or specific gravity low. API-610 has a most useful and widely accepted appendix on recommended configurations for sealing systems.

Metal bellows-type seals for higher-temperature fluids have become more widely accepted. Every major mechanical-seal manufacturer has had considerable experience in designing and applying this seal type for process pumps. Such designs are even available as tandem or double-bellows seals, or with the primary (inner) seal a bellows type, and the secondary or outer seal of the conventional "pusher" type. The latter arrangement will save on first costs, but in high-temperature service provides very short life after failure of the inner seal. An alarm is usually required to indicate inner-seal failure for tandem or double-seal installations. The reader is referred to many excellent supplier application guides and magazine articles [10–15].

On certain small pumps, or on very complex systems on some larger single-stage pumps, the mechanical seals — complete with seal-flushing systems and adequate accessories and controls — can become very expensive. We have actually had examples of small process pumps where the cost of seals plus auxiliaries equaled or approached the price of the pump itself. And, still, the system is subject to eventual failure, or the need for periodic replacement before a catastrophic failure occurs. Perhaps this fact alone has caused considerable interest in leakproof pumps — that is, pumps without seals.

Leakproof pumps

Several types of leakproof pumps have been available from limited sources for many years; the 1980s have seen more pump builders enlarge their lines of pumps to include certain leakproof designs. The two most widely distributed types are the canned-rotor type and the magnetic-drive type.

The canned-rotor type gets its name from the stainless-steel canlike enclosure around the motor rotor, which keeps pumped fluid away from the motor and allows it to return to the pump suction, with or without external cooling. In addition to the familiar single-stage types, several specialized styles are available, including multiple-stage assemblies, vertical inline, high-pressure, self-priming, and even slurry types.

The magnetic drive has its power input from a conventional electric motor to a shaft and bearing arrangement, which, in turn, is connected to a permanent magnet drive within the pump. This results in a leakproof design somewhat less prone to failure because the outside portion of the magnet is thicker than the can in the canned rotor type, and thus less susceptible to corrosion or erosion.

Both types have as their main drawback the inefficiency of the electric motor or magnetic drive, as compared with a conventional pump. However, for completely leakproof pumping, the user may wish to accept this slightly lower efficiency instead of installing very complicated mechanical seals and seal-flushing systems. Table IV shows comparative costs and energy consumptions for conventional, canned-rotor and magnetic-drive pumps. It may not be unreasonable to predict that the next generation of pumps will provide leakproof pumping by means of new sealing methods or with no seals, giving longer service life than designs available today.

Materials for pumps

Notable improvements have been made in pump materials in recent years. Most notable is the increased use of special duplex steel alloys and the newer chrome-nickel alloys. Duplex stainless steels are a combination of ferritic and austenitic stainless steel, a typical one being 26 Cr, 5 Ni 2 Cu, 3 Mo.

Such steels will show qualities superior to the 300-series stainless steels, and have been used for pump casings and internals on seawater, and other difficult services. ASTM A-742 Grade CA6NM is a 13 Cr, 4 Ni-Mo stainless steel that exhibits better casting behavior and is preferred by many pump manufacturers to the older Grade CA-15.

In addition, there is considerable increase in the use of nonmetallic materials for chemical pumps, notably the ANSI type, plus numerous specialized designs, both vertical and horizontal. Nonmetallic centrifugal pumps had until fairly recently been limited to concentric casing designs, for ease of manufacture. But, some are now available with true volute casings, resulting in efficiencies approaching alloy-steel types.

Rotary positive-displacement pumps have also become more widely available in nonmetallic materials for certain parts, such as polytetrafluoroethylene gears within a metal casing, or complete nonmetallic pumps. For controlled-volume pumps, many varieties of materials for primary diaphragms and inlet and outlet assemblies are offered, suitable for handling the most difficult chemicals.

Before specifying some of these specialized materials, the engineer would be well advised to check costs, to find out if the exotic, most-expensive materials are really necessary, or if some less costly, available material would be adequate. Most manufacturers of nonmetallic pumps have very detailed application guides as to just where and where not to use certain materials.

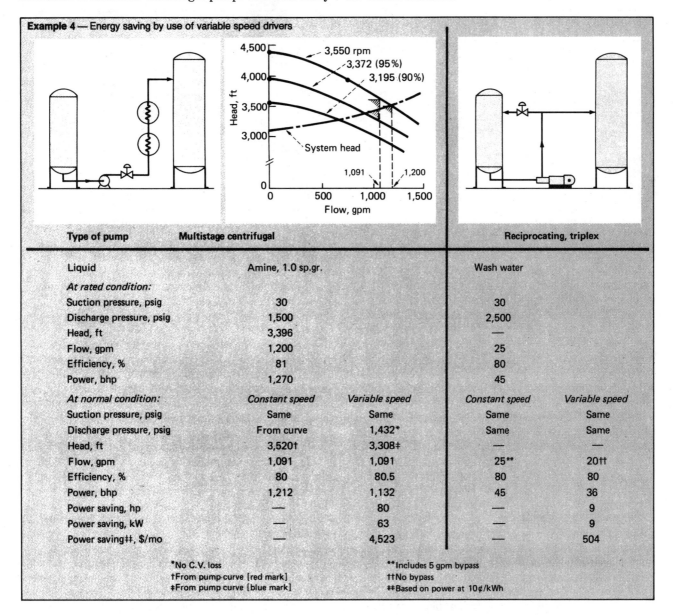

Example 4 — Energy saving by use of variable speed drivers

Type of pump	Multistage centrifugal		Reciprocating, triplex	
Liquid	Amine, 1.0 sp.gr.		Wash water	
At rated condition:				
Suction pressure, psig	30		30	
Discharge pressure, psig	1,500		2,500	
Head, ft	3,396		—	
Flow, gpm	1,200		25	
Efficiency, %	81		80	
Power, bhp	1,270		45	
At normal condition:	Constant speed	Variable speed	Constant speed	Variable speed
Suction pressure, psig	Same	Same	Same	Same
Discharge pressure, psig	From curve	1,432*	Same	Same
Head, ft	3,520†	3,308‡	—	—
Flow, gpm	1,091	1,091	25**	20††
Efficiency, %	80	80.5	80	80
Power, bhp	1,212	1,132	45	36
Power saving, hp	—	80	—	9
Power saving, kW	—	63	—	9
Power saving‡‡, $/mo	—	4,523	—	504

*No C.V. loss
†From pump curve (red mark)
‡From pump curve (blue mark)
**Includes 5 gpm bypass
††No bypass
‡‡Based on power at 10¢/kWh

Advances in positive-displacement pumps

There have been advances in these pumps as well as in centrifugals. Builders of rotary pumps have quickly realized the advantage of the magnetic drive, and many such designs are now available. Such pumps have gained wider acceptance in metered flow services, where the exact accuracy of controlled-volume pumps may not be required. For small-flow process requirements (below 20 gpm) users should continue to look at rotary pumps, which are usually more efficient than small centrifugals and are available in almost any required style and metallurgy.

Controlled-volume pumps have advanced significantly in their ability to provide leakproof pumping through the wider use of diaphragm-type liquid ends. Such designs are now available in larger sizes or for higher pressures, or both, and will provide safe means for pumping very toxic, dangerous materials. Alarms are arranged in double-diaphragm pumps

Table IV — Leak resistance, not efficiency, favors canned-rotor and magnetic-drive pumps over conventional ones

	Conventional process pump, 3 × 4 × 8½	Canned-rotor pump, 3 × 4 × 10	Magnetic-drive pump, 3 × 4 × 8½
Speed, rpm	3,500	3,450	3,500
Efficiency, %	65	57	65
Pump hp	30.6	34.8	30.6
Magnet losses	None	None	4.5
Total bhp	30.6	34.8	35.1
Motor size	40	40	50
Motor eff., assumed	0.92	0.88	0.92
kW required	24.8	32	28.5
Power cost, $*	19,840	25,600	22,800
Initial cost factor, for pump and motor	1.0	2.0 – 2.5	2.0 – 3.0

*at 10¢/kWh, 8,000 h

to warn of the failure of the process diaphragm and attract the operator's immediate attention. Diaphragm designs have been successfully applied at pressures to 3,500 psi and higher; also, the tubular-diaphragm type is available with similar leakproof features in sizes to 400 gal/h or pressures to at least 3,500 psi.

Other diaphragm pumps have found more applications as the need has increased for pumps to handle slurries or other troublesome mixtures. Air-operated diaphragm pumps are widely used for sump or pumpout service, difficult slurry service, and in the construction industry. Recently, several large-size plunger-type reciprocating pumps, using multiple-diaphragm-type liquid ends have been installed for pumping red mud in alumina-ore processing, where the pumping rate is over 1,300 gpm at 1,100-psi differential pressure, absorbing more than 1,200 bhp. These may represent the largest diaphragm-type pumps to have been built up until now. Other reciprocating diaphragm pumps have been built for pressures to 50,000 psi.

In the 1980s, pump users have become more aware of the need to specify NPSH requirements on positive-displacement pumps, and to consider carefully the effect of the acceleration head in the system design. Example 5 shows a simple system employing a small, triplex, plunger pump, and illustrates how specialized this pump selection became when the actual pump chosen for the job was a conventional plunger pump, oversized and running at about 50% of maximum speed to meet the NPSHA conditions.

Newer types of pumps

Other new types of pumps have appeared and gained acceptance. The pitot tube design is a very-low-specific-speed type (N_{ss} = 80 to 250 in U.S. units). It is capable of low-flow, high-head duty with a dynamic-type design. A rotating inner casing contains the pumped fluid, and pressure is increased as the fluid passes a stationary pitot tube within the pump. Speeds range from 1,750 to 5,400 rpm in currently manufactured models.

The disk pump has gained some popularity in pumping special materials including viscous fluids, shear-sensitive materials such as latex or polymer emulsions, non-Newtonian fluids, slurries and many others. Flow range is to 3,000

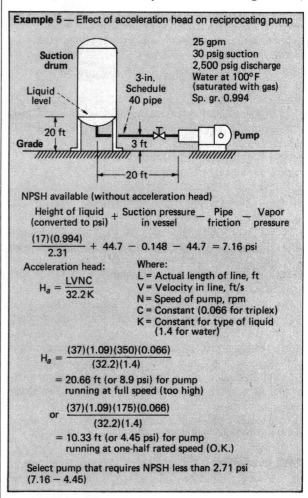

Example 5 — Effect of acceleration head on reciprocating pump

25 gpm
30 psig suction
2,500 psig discharge
Water at 100°F (saturated with gas)
Sp. gr. 0.994

NPSH available (without acceleration head)

Height of liquid (converted to psi) + Suction pressure in vessel − Pipe friction − Vapor pressure

$$\frac{(17)(0.994)}{2.31} + 44.7 - 0.148 - 44.7 = 7.16 \text{ psi}$$

Acceleration head:

$$H_a = \frac{LVNC}{32.2K}$$

Where:
L = Actual length of line, ft
V = Velocity in line, ft/s
N = Speed of pump, rpm
C = Constant (0.066 for triplex)
K = Constant for type of liquid (1.4 for water)

$$H_a = \frac{(37)(1.09)(350)(0.066)}{(32.2)(1.4)}$$

= 20.66 ft (or 8.9 psi) for pump running at full speed (too high)

or $\frac{(37)(1.09)(175)(0.066)}{(32.2)(1.4)}$

= 10.33 ft (or 4.45 psi) for pump running at one-half rated speed (O.K.)

Select pump that requires NPSH less than 2.71 psi (7.16 − 4.45)

gpm, heads to 500 ft and N_s to about 2,000. In this pump, the rotor is a series of discs on one shaft, and energy is imparted through the principles of boundary layer and viscous drag, following basic fluid-mechanics theory. Another, even newer, design, consisting of a nutating disc, has appeared and may be applicable for specialized applications.

An upgraded medium-duty chemical-process centrifugal pump has been proposed [17], and is available from at least one U.S. source. In this pump, some of the basic ideas from the ANSI B73 Standards are retained, but the pump is not dimensionally interchangeable, size for size, with B73 types. Bearings, shafts and seals are heavier, and the proponents of this pump feel it fills a need in a range intermediate between present ANSI B73 types and the API-610 type. This writer (and others [18]) do not feel that the UMD pump is necessary.

We should prefer to retain the modern versions of the ANSI B73 pumps, and encourage manufacturers to give attention to some of the newer subjects addressed in this article.

Reference standards for pumps

In the U.S., the major industrywide pump standards for application and specification continue to be the Hydraulic Institute Standards (14th Ed., 1983), the ANSI Standards for chemical pumps — B73.1 (horizontal type) and B73.2 (vertical inline type) — plus the American Petroleum Institute Standard API-610 (centrifugal pumps), API-674 (reciprocating pumps), API-675 (controlled-volume pumps) and API-676 (rotary positive-displacement pumps). API-610 is currently in its sixth edition (1981); the seventh edition has been completed by an API taskforce and will hopefully be published in 1988.

ASME is considering the need for updating ANSI B73.1 and B73.2. Also, ASME is preparing "ASME Design Rules for Pumps (including Proposed Shaft Design Rules)," and ASME Power Test Codes for pumps are still widely referenced and may be in for some updating [19].

Worldwide, an ISO (International Organization for Standardization) publication has been written in England and elsewhere in Europe, covering "medium duty" centrifugal pumps, somewhat more detailed than the ANSI B73.1 and .2, and more specifically aimed at medium-duty single-stage process pumps. All of the above standards are for use as as guides; most users and larger constructors will have their own standard specifications, often based on the ANSI or API Standards.

Several useful new books have appeared in the 1980s relating to pumps (see references at the end of the article). Use of pump data sheets as a means of specifying pumps for each individual service continues to be the most common procedure. In the very near future, we can hope to see, the use of computer-aided techniques for preparing and completing such data, which it is hoped will speed up the tedious task of preparing pump specifications and permit the engineer of the future to devote more attention to the total system needs and the careful integration of each pump into the process.

Testing of pumps

Users are specifying more pump testing than previously. This is especially true for larger, complex, custom-designed pumps, for which elaborate, detailed performance testing is specified and witnessed by the user or the user's representatives. Manufacturers have accommodated user needs by enlarging pump-test capabilities and recording of data in more-accurate, up-to-date methods. Use of computers has become the industry standard for accepting the data and reducing them into final output form.

Nondestructive testing and examination of critical pump parts, such as large castings and forgings, has likewise become a necessity in the opinion of most large users; the risk of delay on long-lead-time parts, or of unexpected failures upon plant startup, has caused more users to insist that the quality-control and quality-assurance methods of pump manufacturers are carefully monitored. Even third-party inspection (i.e., manufacturer, purchaser, plus an independent third party such as Bureau Veritas or Lloyd's of London) has become quite common for complex castings, forgings and pump tests.

Conclusion

Let us reemphasize that pump technology has advanced rapidly in recent years, and that the users and builders of pumps can expect further advances over the years.

References

1. "Cost Oriented Systems Technique," ICARUS Corp., Rockville, Md.
2. Corripio, A. B., others, Estimate Costs of Centrifugal Pumps and Electric Motors, *Chem. Eng.*, Feb. 22, 1982, p. 115.
3. Fraser, Warren H., Recirculation in Centrifugal Pumps, ASME Annual Meeting, Nov. 1981.
4. *Power and Fluids*, Vol. 8, No. 2, 1982 (pub. by Worthington Group, McGraw Edison Co.).
5. Karassik, I. J., Pump Operation at Off-Design Conditions, a three-part series, *Chem. Process.*, 1987.
6. Trans. Third Intel. Pump Symp., May 1986, pub. by Texas A&M University, College Station, Tex.
7. Gopalakrishnan, S., "Minimum Flow Criteria," Byron Jackson Pump Div., Borg-Warner Corp., Los Angeles, Calif., 1986.
8. Taylor, I., Pump Bypasses, Now More Important, *Chem. Eng.*, May 11, 1987, p. 53.
9. Doll, Thomas R., Making the Proper Choice of Adjustable Speed Drives, *Chem. Eng.*, Aug. 9, 1982, p. 46.
10. "Mechanical Seal Manual," B-W Seals, Inc., Temecula, Calif.
11. "John Crane Mechanical Seal Manual" John Crane, Houdaille Inc., Chicago, Ill.
12. "Mechanical Seal Manual," Durametallic Corp., Kalamazoo, Mich.
13. "Mechanical Seal Handbook," Sealol Div., EG&G Inc., Warwick, R.I.
14. Meyer, E., "Mechanical Seals," 5th ed., (Burgemann Co.), Butterworth & Co., London, 1977.
15. Ramsden, J., How to Choose and Install Mechanical Seals, *Chem. Eng.*, Oct. 8, 1978, p. 97.
16. Henshaw, T. L., Reciprocating Pumps, *Chem. Eng.*, Sept. 21, 1981. p. 105.
17. Bloch, H. P., and Johnson, D. A., Downtime Prompts Upgrading of Centrifugal Pumps, *Chem. Eng.*, Nov. 25, 1985, p. 35.
18. Reynolds, J. A., Standard Pumps Are Not Obsolete!, *Chem. Eng.*, May 12, 1986, p. 119.
19. ASME Power Test Codes: PIC-7.1, Displacement Pumps; PIC-8.2, Centrifugal Pumps.

Recommended additional references

Dicmas, J. L., "Vertical Turbine, Mixed Flow, and Propeller Pumps," McGraw-Hill, New York, 1987.
Karassik, I. J., others, "Pump Handbook," 2nd ed., McGraw-Hill, N.Y., 1986.
Chemical Engineering Guide to Pumps — pub. by *Chemical Engineering*, McGraw-Hill Publications Co., New York, 1984.
"Hydraulic Institute Standards," 14th ed., pub. by the Hydraulic Institute, 712 Lakewood Center North, 14600 Detroit Ave., Cleveland, OH 44107, 1983.
Lobanoff V. S., and Ross, R. R., "Centrifugal Pumps: Design and Application," Gulf Pub. Co., Book Div., Houston, Tex., 1985.
"Cameron Hydraulic Data," 15th ed., Ingersoll-Rand Co., Woodcliff Lake, New Jersey, 1977.
Karassik, I. J., and Carter, R., "Centrifugal Pumps," McGraw-Hill, New York, 1960.
Stepanoff, A. J., "Pumps and Blowers: Selected Advanced Topics," 1st ed., John Wiley & Sons, New York, 1965.
Stepanoff, A. J., "Centrifugal and Axial Flow Pumps: Theory, Design and Application," 2nd ed., John Wiley & Sons, New York, 1957.
API Pump Standards 610, 674, 675, 676, American Petroleum Institute, Washington, D.C.
ANSI Pump Standards B73.1 and B73.2, American National Standards Institute, New York, N.Y.

The author

Richard F. Neerken is section manager of process equipment for The Ralph M. Parsons Co., 100 West Walnut St., Pasadena, CA, 91124 (telephone: (818) 440-3498), where he directs a group of more than 30 engineers doing application engineering on rotating machinery, including pumps, turbines, compressors and engines. He holds a B.S. in mechanical engineering from California Institute of Technology, is a registered professional engineer in California, a member of the American Soc. of Mechanical Engineers, and a member of the Contractors Subcommittee on Mechanical Equipment for the American Petroleum Institute.

New pump models aim at satisfying CPI needs

Units that avoid leakage have gained in popularity since the Bhopal tragedy, but demand is also great for plastic pumps for corrosive service and pumps that save energy.

The ideal pump should probably be leakproof, energy-efficient, maintenance-free (but easy to fix if something goes wrong), completely resistant to whatever corrosion and abrasion conditions it was designed for, and at the most no more expensive than the last pump the purchaser bought. These are the qualities that customers want, say U.S. pump manufacturers, who are trying hard to meet requirements that may rarely come together in any one product.

WHAT'S SELLING?—Demand for leakproof pumps is being reflected in a thriving business for makers of seal-less pumps. "Our sales have increased by an average of 25% a year for the past 10 years. But in 1984, sales were double those of 1983," says Jan Hatch, marketing manager for The Kontro Co., Inc. (Orange, Mass.), a firm that specializes in magnetically-driven pumps.

Other manufacturers note that, while the demand for better seals and seal-less pumps has been building for a long time, it has intensified since the Bhopal incident. "Our backlog has quadrupled since then," says Dennis Fegan, sales manager for canned-motor pumps at the Pacific Pumps Div. of Dresser Industries Inc. (Fort Washington, Pa.).

In another approach to the leakage problem, manufacturers of other types of units are supplying or developing improved seals. Also, sales of cartridge seals are reportedly increasing, despite their higher cost, because they avoid

Originally published April 1, 1985

Technician dissembles a centrifugal pump that has oversized seals

the hazard of faulty installation associated with conventional mechanical seals, and feature easier maintenance and adjustment.

Sales of reinforced-plastic pumps are rising also, because they resist corrosion and cost much less than the special alloys or exotic metals that might otherwise have to be used.

The push for energy efficiency is evident in the increasing popularity of variable-frequency motors to vary pump speed in response to flow changes. Pumps are also being designed and manufactured to be more efficient, and producers are making more use of investment casting, which — although more expensive — results in more-precise dimensions and a smoother finish than traditional sand casting.

"A major need is to improve the efficiency of smaller centrifugal pumps," says Alexander Agostinelli, director of product planning for the Worthington Div. of Dresser Industries (Basking Ridge, N.J.). "About 90% of the pumps sold are 20 hp or less (i.e., ranging up to about 3-in. discharge size), and these are only about 45–55% efficient." Worthington is researching improved designs and manufacturing techniques.

SEAL TALK—A pump designed with the idea of satisfying most requirements at a lower cost is being offered by A. W. Chesterton Co. (Stoneham, Mass.), a newcomer to the pump business. The firm, which has mechanical-seal experience, made a study that determined that 80–90% of pump shutdowns are due to seal or bearing failures. Chesterton then used its seal know-how to design a centrifugal pump around oversized seals that provide more room for cooling, flushing and cleaning, plus an oversized bearing chamber that carries extra lubricant and an oil filter.

Modular pump parts are interchangeable among the 16 pump sizes. Chesterton also sells a handheld analyzer to measure pump vibration, temperature and pressure. (For more details, see *Chem. Eng.*, Feb. 20, 1984, p. 53.)

"The problem with most pumps is that the space for mechanical seals is very tight because it was originally made for packing, but pump manufacturers don't want to change their designs," says Richard Ankoviak, as-

NEW PUMP MODELS AIM AT SATISFYING CPI NEEDS

sistant manager of Chesterton's Advanced Systems Div. He admits that the pump is about twice the price of a standard one, but claims that it more than pays for itself because the seals and bearings last twice as long.

Established pump manufacturers are skeptical of these claims, and say that there have been problems with the pump. "It's a gimmick," asserts H. Russell Young, corporate marketing manager for Goulds Pumps, Inc. (Seneca Falls, N.Y.). "We wouldn't hesitate to change our pump design to provide a better mechanical seal."

Robert Cerza, a senior applications engineer with Goulds, and a member of the ANSI B73 committee, adds that the B73 standard (for pumps) is adequate because it specifies that the stuffing box be designed for both packing and mechanical seals.

However, the provision of better seals is a major concern of many pump manufacturers. "We are working with seal makers to develop better products, and we would redesign our stuffing boxes if we had to," says Richard Andrews, marketing manager with Ingersoll-Rand Co.'s Standard Pump Div. (Allentown. Pa).

A. R. Wilfley & Sons, Inc. (Denver), whose claim to fame is a packingless centrifugal pump that uses a rotating expeller in place of a seal, is just putting on the market an improved design that has a diaphragm seal that closes when the pump stops.

Wilfley's standard pumps also have a seal that does that, but it is more complex than the new one. When the pump is working, sealing is effected by an expeller (or expellers) located on the shaft behind the impeller. The expeller exerts a centrifugal force to keep liquid away from the shaft; it prevents leakage, even though the rear of the pump is open to the atmosphere. When the pump stops, that opening is closed by a mechanical seal, activated by two spring-loaded governor weights attached to the shaft sleeve. The weights fly out when the shaft rotates, then fold down on the sleeve when it stops.

The new design, called the Model ES acid-sludge pump, has a similar configuration, except that the seal is replaced by a flexible, doughnut-shaped elastomer diaphragm attached to a ring. During pumping, the diaphragm is in the open position and does not rub on the sealing surface (the end of the shaft sleeve). When the pump slows to a stop, the expellers lose their effect and pressure builds up, causing the seal to close.

"This seal is just as reliable as our older system, but it is less complicated and less expensive," says William Wilbur, marketing communications manager. The pump is offered in hard iron, elastomer, stainless steel, or a combination of those materials in 4x3 (intake x discharge), 6x4, and 8x6-in. sizes.

Goulds has also come out with a dynamic seal that works on the same principle as Wilfley's expeller. It is located behind the impeller, and has straight radial vanes that repel liquid that gets past the impeller. Goulds offers it as an option on its entire line of 3196 chemical pumps and 3175 paper-industry pumps.

"We see a growing market for this type of seal," says Young. "It adds 25–40% to the price, but it doesn't leak when the pump is working, and it has

---

"When a company needs a seal-less pump, efficiency isn't a major influence in the buying decision."

---

two to four times the life of a standard mechanical seal." Goulds also has a kit that permits customers to convert existing pumps to accommodate the new seal.

The Duriron Co.'s Pump Div. (Dayton, Ohio), which already offers an expeller seal on some centrifugal pumps, will introduce this month a new line of fiberglass-reinforced epoxy pumps featuring a new mechanical seal. Essentially, it reverses the positions of the conventional seal elements. "What one might consider the stationary seal will be the spinning part in our case, and the flexible parts will be stationary," explains Glenn George, new-products manager. "The significance is that the low-mass piece spins rather than the high-mass part, and this avoids wear and tear when wobble occurs." The new design is expected to reduce seal failures and lengthen seal and shaft life.

SEAL-LESS MODELS—Seal-less pumps (i.e., magnetically-driven and canned-motor pumps) have a reputation for inefficiency, and are more expensive than conventional centrifugal pumps, but the need to contain hazardous, toxic or expensive chemicals has offset such considerations. And seal-less-pump makers say efficiencies are improving.

"When a company needs a seal-less pump, efficiency isn't a major influence in the buying decision," says Dean Werner, vice-president of sales and marketing for LaBour Pump Co. (Elkhart, Ind.). LaBour's pumps use permanent magnets (some types have electromagnets).

Werner claims that LaBour's newest pump, introduced last fall, is 45–60% efficient, which is high for a magnet-drive pump. It has a metal body and a liner of ethylene tetrafluoroethylene or polyvinylidene fluoride. The company prefers metal for more-rugged bodies, although many seal-less pumps are made of reinforced plastic. The new pump line has a maximum capacity of 10 hp and 325 gal/min.

Kontro avers that the efficiency of its magnet-drive pumps is approaching 90% of that of conventional centrifugal pumps, up from about 55% several years ago. The reason is that the magnet ring inside the pump is connected directly to the shaft behind the impeller, says Hatch.

"The torque capability of our magnetic drives is far greater than those of other magnet-drive pumps," he asserts. Kontro uses powerful permanent magnets in metal-bodied pumps of up to 500 hp, compared with a maximum of about 15 hp for typical magnet-drive units. The company last year introduced what it claims is the first seal-less pump to meet the American Petroleum Institute 610 standard (*Chem. Eng.*, Dec. 24, 1984, p. 39). More recently, Kontro claimed another "first" — a seal-less self-priming pump with a Hastelloy body that can withstand a vacuum without deformation (in search of efficiency, seal-less pump bodies usually are very thin).

In May, Apex Chemical Equipment Co. (Elk Grove Village, Ill.) plans to introduce what it claims is the first magnet-drive pump to meet ANSI standards. The fiberglass-reinforced fluoroplastics unit is designed to resist hydrochloric and hydrofluoric acids, and will be available with 2–10-hp motors.

As for cost, Hatch says that Kontro's pumps are about 25% more than conventional ones. Fegan, of Pacific Pumps, notes that while the initial expense of a canned-motor pump can be double that of a conventional unit, the installed cost can be "quite close," because of the cost of seals and of aligning the conventional pump with a motor.

Because of the market growth, more manufacturers are entering or planning to enter the market. Sundstrand Fluid Handling (Arvada, Colo.) started making canned-motor pumps last year under license from Japan's Nikkiso Co. Others researching the market include Ingersoll-Rand and Goulds.

MORE PLASTIC PUMPS—As the market for nonmetallic (essentially, reinforced plastic) pumps expands, the major effort is to make equipment that can withstand higher temperatures without deformation. For instance, Duriron's new line of fiberglass-reinforced plastic (FRP) pumps is designed for temperatures up to 300°F, compared with 225°F for its earlier FRP models.

"We got the higher temperature capability mostly through modifications and additives," says George, noting that the company makes its own molding compounds. Besides the all-FRP pump, Duriron will have a new perfluoroalkoxy (Teflon)-lined product, also good for 300°F operation, and a 225°F vinyl ester line for bleach.

At about mid-year, Worthington plans to introduce its first line of FRP centrifugal pumps (up to about 8-in. discharge) for lower-temperature use. Goulds also intends to start marketing FRP centrifugal pumps, but toward the end of the year. "We hope to sign a licensing agreement very soon to use a state-of-the-art molding process," says Goulds' Young.

Warren Pumps-Houdaille Inc. (Warren, Mass.) already makes reinforced-plastic pumps with 300°F capability, and is shooting for 400°F by yearend, says James Paugh, vice-president of engineering.

Fybroc Div. of Met-Pro Corp. (Hatfield, Pa.) has just introduced what it claims is the first full line of FRP self-priming pumps. The five sizes range from 3x2x6 in. to 6x4x13 in., and handle up to 1,200 gal/min. They are available in either vinyl ester or epoxy, the two most commonly used classes of materials.

Charles Drake, Fybroc's manager of sales and marketing, explains that vinyl esters are generally better for acids and caustic, but are limited to around 220°F, while epoxies are better for solvents and have higher temperature capability.

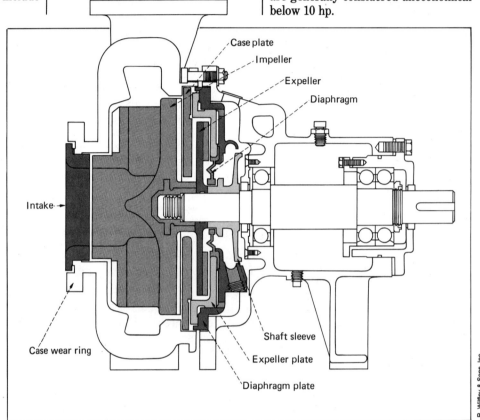

New pump model has a diaphragm seal that works when the pump stops

George, of Duriron, notes that a number of plastic materials can handle higher temperatures, but the trick is to find plastics that combine this with good chemical resistance, mechanical properties, processability and reasonable cost. "Some years ago we tried to make a pump of phenolic, which is good for chlorinated solvents, but it didn't have good mechanical properties."

Glass can be an "Achilles heel" in an FRP pump, says Edward Margus, vice-president of engineering for Vanton Pump & Equipment Corp. (Hillside, N.J.). He points out that it can be attacked by hydrofluoric acid and fluoride compounds if it is exposed at the surface, and can also serve as a path into the plastic (for chemicals). Vanton makes centrifugal, rotary and vertical sump pumps of various materials, including vinyl and propylene, but uses no reinforcement (except for the plastic-covered steel shafts).

CONTROLLING THE FLOW—Variable-frequency drives are finding increasing use in saving energy and avoiding pump wear when less flow is required, say manufacturers, but at present they are generally considered uneconomical below 10 hp.

The Byron Jackson Pump Div. of Borg-Warner Industrial Products, Inc. (Long Beach, Calif.) offers an alternative to variable-frequency motors: modification of installed pumps to reduce their flow capacity. The company replaces the impeller with a specially-designed smaller one, and also installs a volute or diffuser and oversized wear rings to reduce the flow.

"We have been doing this for about 18 months and it has been very successful," says Jeffrey Hohman, manager of business development. "A lot of these pumps were overdesigned in the first place, and we have found many cases where they are being throttled to about 50% of capacity by tightening up the discharge-control valve. In these cases

NEW PUMP MODELS AIM AT SATISFYING CPI NEEDS

we have saved up to $50,000 a year in energy costs."

Byron Jackson has designed four different sizes of various parts that can be used to modify 40 different pump sizes, says Hohman. He adds that the company is considering employing this modular approach in the design of a new pump, with the idea of saving in manufacturing costs and spare parts.

High efficiency is claimed for a new multistage centrifugal pump introduced last year by Union Pump Co. (Battle Creek, Mich.). The first five units were sold for the difficult job of pumping liquid carbon dioxide, says a company spokesman, adding that efficiency averaged about 75%. He attributes this to careful design and the use of investment casting.

Milton Roy Co. (Ivyland, Pa.) is working privately with an a.c.-drive company to develop an efficient means of employing a variable-frequency motor on its metering pumps, which use a hydraulically-actuated diaphragm operated by a piston.

"We have tried an off-the-shelf a.c. controller to vary the stroke speed and it doesn't work," says Jack Kelley, marketing, and research and development manager for the flow control division. "The problem is that the controllers can't recognize the variable torque, and if the motor keeps slipping because of this, the error in metering becomes substantial." Milton Roy hopes to resolve this by replacing the complex electromechanical linkage with a simpler system that will deliver a varying stroke speed.

Kelley notes that there is a general trend away from plungers and toward diaphragms, which have more universal use. However, he says plungers have found a market in new high-pressure applications (up to around 6,000 psi), such as supercritical extraction, and polymer injection for enhanced oil recovery.

"More pump for the buck" by going to higher stroke speeds is a trend observed by Robert McCabe, manager of product engineering for metering pumps with Pulsafeeder Div. of Clevepak Corp. (Rochester, N.Y.). In the past two years, Pulsafeeder has increased the stroke speed of its diaphragm pumps from 140 to 175 strokes/min, which is typical, he says, adding that there are some on the market above 200 strokes/min.

"Compressed air is getting too expensive, so I think a lot of people are looking at other methods, such as hydraulics or an electric motor with an eccentric drive."

MAKING CHANGES—Viking Pump-Houdaille Inc. (Cedar Falls, Iowa) last year modified a line of abrasive-liquid rotary pumps to achieve about 30% more throughput. Among the changes: the substitution of silicon carbide for ceramic facings to avoid overheating at higher speeds, and of tougher tungsten carbide for hardened steel in gear parts. Tuthill Corp.'s Pump Div. (Chicago) uses self-lubricating carbon bearings in a new internal-gear circumferential piston pump, which allows the unit to move nonlubricative fluids.

Wilden Pump & Engineering Co. (Colton, Calif.) has cut energy costs 10% in a new model of its air-operated double-diaphragm pump by redesigning it so that less air pressure is needed.

Roper Pump Co. (Commerce, Ga.) also is seeking ways to cut the energy consumption of its double-diaphragm pumps. "Compressed air is getting too expensive, so I think a lot of people are looking at other methods, such as hydraulics or an electric motor with an eccentric drive," says Vern Landwehr, the firm's product sales manager.

Nash Engineering Co. (Norwalk, Conn.) last year introduced a new line of large liquid-ring vacuum pumps (4,000–12,000 ft^3/min) that cut by 50% the amount of water needed to compress the gas for the pump. The company is now changing its smaller pumps to the new design. Nash has a patented water recirculation device that reduces water consumption from 150 gal/min to 75 gal/min in the large pumps. The new design also reduces energy consumption by 10–15%.

Sullair Corp. (Michigan City, Ind.) says that its recently introduced rotary-screw vacuum pump is more efficient than its liquid-ring vacuum pumps for higher vacuums. "It is about equal at 20 in. of mercury, but at least twice as efficient at 29 in.," says Wayne Benson, general manager of Sullair's Vacuum Div. However, he says it is unsuitable for dirty gas streams because it does not have the scrubbing action of a liquid-ring pump.

Manufacturers of liquid-ring vacuum pumps say one of their biggest markets is the replacement of steam ejectors.

MISCELLANEOUS ENTRIES—A novel centrifugal pump that uses flat disks rather than impellers is finding a market in rugged applications, such as pumping abrasive slurry, according to the manufacturer, Discflo Corp. (Santee, Calif.).

"We have sold about 100 so far," says Edward Korbel, president. Each of the parallel disks has an inlet hole in the center. The pump works on the principle of boundary-layer drag (like that around an airfoil), creating friction that moves material through.

Cost is about 15% more than a standard centrifugal model, says Korbel, but maintenance costs are much less, especially in pumping abrasive slurries, where the flat-disk unit lasts much longer. He adds that the pump is also being used successfully to move sensitive materials, such as plastic resin, because its action does not harm the product.

Discflo is less than three years old, but Max Gurth, the executive vice-president, previously tried to market the pump through another company, Eureka Pumps International, Inc., of which he was president.

Also marketing a disk pump is U.S. Pump & Turbine Co., Paramount, Calif. Formed last summer, the company bought license and patent rights from General Enertech, whose president, Jake Possell, spent a number of years working on disk pumps and made a number of test installations. Both Possell and Gurth have patents and conflicting claims over the concept.

Another novel pump—a submersible design with a double-fluted impeller that works rather like an extruder—has been tested successfully to move peas, broccoli and other foods in a process plant, according to its developer, Harold Owens, president of Heli-Arc and Manufacturing Co., a machine and welding jobshop in Twin Falls, Idaho. "The food industry needs a pump to handle food gently," he says, explaining that his screw-shaped impeller works with neutral pressure to push product through, "rather than sucking it up and pushing it out." Owens plans to market his pump to the food industry.

Part II
Centrifugal pumps

Downtime prompts upgrading of centrifugal pumps
Standard pumps are *not* obsolete!
Head-vs.-capacity characteristics of centrifugal pumps
Multistage centrifugal pumps
Time takes its toll on centrifugal pumps
Designing centrifugal pump systems
Pump bypasses now more important
Centrifugal pumps and system hydraulics
Unusual problems with centrifual pumps
Startup of centrifugal pumps in flashing or cryogenic liquid service
Choosing plastic pumps

DOWNTIME PROMPTS UPGRADING OF CENTRIFUGAL PUMPS

Heinz P. Bloch, Exxon Chemical Co., and **Donald A. Johnson,** Carver Pump Co.

Seven improvements in the design of medium-duty process pumps could lengthen mean time between failures from the average of 13 months for existing pumps to 25 months. Hydraulic performance — hence, energy efficiency — could be enhanced by the elimination of present dimensional constraints. The design changes needed were identified by the application of reliability-engineering concepts.

The vast majority of pumps in chemical process plants and oil refineries operate below 300 psig and 350°F. Even at these conditions, an inordinately high percentage of the maintenance expense for rotating equipment goes into pump repairs. A typical work order for pump repair in a Gulf Coast process plant exceeds $5,000. The reduction of such repair costs through the improvement of pump reliability should, obviously, rank high on any list of plant priorities.

Computerized record-keeping has made it possible to identify when and why a particular piece of equipment failed, and to compare its failure frequency and repair costs with those of other such equipment operating under similar process conditions. Such record-keeping has revealed the following:

• Standard ANSI (American National Standards Institute) pumps have a mean time between failures of only 13 months even in well-maintained facilities in North America; the industry average is closer to six months.

• Attempts to correct the causes of pump failures have traditionally focused on organizational approaches. This is no longer enough; it is time to look to pump manufacturers for fundamental improvements.

• More than 40% of the ANSI pumps now installed can be replaced with types having an average efficiency that is often 10% higher, or even more.

• API pumps in medium-duty service are usually not cost-effective.

These findings made it apparent that deficiencies of the ANSI pump needed to be identified. An Exxon Chemical Co. study (see box, p. 24) revealed the following weaknesses: (1) shaft deflection is excessive, (2) the dimensional limits imposed by ANSI B73.1 do not allow sufficient space for the application of the superior mechanical seals now available and often curtail the attainment of optimum efficiency, (3) bearing life is shorter than it could be because of shortcomings in present bearing designs and lubrication systems, (4) frangible pressure-containment holding devices create an unnecessary safety hazard, and (5) the ANSI-design baseplate does not provide adequate structural integrity and load-bearing capability.

Excessive shaft deflection

Pumps of overhung impeller construction are prone to excessive shaft deflection, which leads to internal wearing of wear rings, bushings and sleeves, and reduces the life expectancy of mechanical seals. In many instances, frequent stress reversals cause shaft fatigue. Because operating and maintenance costs tend to rise with increasing shaft deflection, reduced deflection should be a key feature of a new pump design.

The amount of shaft deflection can be calculated via the following equation [1]:

$$Y = \frac{F}{3E}\left(\frac{N^3}{I_N} + \frac{M^3 - N^3}{I_M} + \frac{L^3 - M^3}{I_L} + \frac{L^2 X}{I_X}\right)$$

Here, Y = shaft deflection at impeller centerline, in.; F = hydraulic radial unbalance, lb; M and N = distances from impeller centerline to the steps on the shaft, in.; L = distance from impeller centerline to centerline of inboard bearing, in.; X = span between bearing centerlines, in.; I_L, I_M, I_N and I_X = moments of inertia of the various diameters, in^4; and E = modulus of elasticity of shaft material, psi (see Fig. 1)

For a reasonably accurate approximation, let $Y = FL^3/3EI_M$. Note that, because $I_M = \pi D_M^4/64$, shaft deflection is a function of L^3/D^4. At least one major oil company makes use of this fact in its engineering specifications. Its pump data sheets include spaces for the L^3/D^4 ratio to be provided by the vendor. This ratio is called the shaft flexibility factor, or SFF. In competitive bidding, the SFF values given by bidders are compared against the lowest SFF value, and the higher values are assigned a "maintenance assessment" of a certain percentage of their bid price (see Fig. 2).

Shaft deflection changes as a function of the fluid flowrate through the pump. As a pump's capacity increases or decreases, moving away from the best efficiency point, the pressures around the impeller become unequal, tending to deflect it. In an overhung impeller pump having a standard single volute casing, this deflection can reach serious magnitudes (Fig. 3). Special casings — such as diffusers and double-volute and concentric casings — can greatly reduce the radial thrust and, hence, the deflection. However, not even the best-designed casing can completely eliminate pressure-induced shaft deflections.

That a single-volute pump will not give satisfactory long-term service if operated too far from its best efficiency point is indicated by Fig. 3. Regardless of the volute design, the mechanical strength of a pump tends to be improved by low shaft flexibility factors.

Dimensional limits imposed by ANSI B73.1

Conventional pumps have casing back covers that incorporate integrally-cast stuffing boxes. Stuffing-box size limitations and general dimensional constraints have for years impeded the application of advanced high-performance mechanical face seals in many overhung ANSI-type pumps.

Subsequent to a fluid-sealing forum, sponsored by the fluid sealing industry, at which the inadequacy of present pump seal boxes in accommodating newer mechanical seals was discussed [2], symmetrical packaged seals (Fig. 4) became available. Packaged assemblies of this type having single, double or tandem mechanical seals can now be purchased [3]. Such a package can completely replace the stuffing box of many pumps whose hydraulic ends are adequate but whose stuffing-box dimensions are unsuitable for reliable long-term seal operation or for the installation of the better-designed seals.

Suitably configured seal packages can be procured from seal manufacturers. The symmetrical packaged seal or a unitized seal can be mounted to the pump casing by the seal manufacturer, or this can be done by the pump manufacturer, in accordance with the seal manufacturer's installation instructions.

These newer seals open up possibilities for significantly improving the reliability and reducing the operating cost of a wide spectrum of centrifugal pumps.

For economic or technical reasons, there may be instances when it is best to incorporate in a pump a stuffing box that

Company study disclosed need for improved centrifugal pump

In 1983, Exxon Chemical Co. initiated a program to interest pump manufacturers in modifying or redesigning existing ANSI-type chemical process pumps in order to reduce operating and maintenance costs. The pumps of concern were foot-mounted medium-duty chemical process pumps that operate at up to 300 psig at 350°F. Exxon required that the new designs lead to improved mean time between failure (MTBF) — a factor related to total pump life-cycle cost in a complete return-on-investment analysis.

The new pumps' range of operating curves was to be consistent with the present ANSI B73.1 range of hydraulic performance (head and capacity) coverage, although compliance with the dimensional standards of ANSI B73.1 was judged not to be of great importance.

Realizing the need for new pump specifications, Exxon initiated an evaluation of the centrifugal pumps in its operating units. Its findings supported Carver's own research that the time had come to offer the chemical process industries a cost-effective alternative to the ANSI and API designs based on life-cycle-cost analysis.

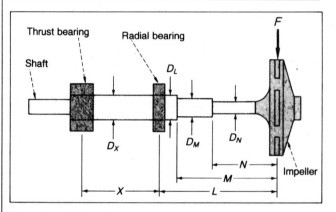

Figure 1 — Shaft deflection depends on these dimensions

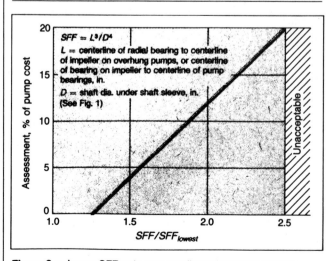

Figure 2 — Lower SFF ratio gets smaller assessment penalty

Maintenance vs. capital — What does a pump actually cost?

Many operating companies and engineering firms regard chemical process pumps as a commodity purchase item. Most often, a basic engineering specification for an ANSI pump is sent to the purchasing department, which gives the order to the lowest bidder. Little or no thought is given to:
- What the maintenance cost of the pump will be over its life cycle.
- What the cost of power will be for operating the pump, compared with the cost of running the most efficient pump now available.
- What the life-cycle cost of the pump will be in its intended service.

Only when the foregoing factors are considered does the actual cost of a pump become apparent. For example, repair costs can easily exceed the price of a new pump. Even so, better remedies for pump problems have been resisted in some companies because a departure from established standards would require changing piping and replacing spare-parts inventories, and because of the comfort of sticking with ANSI standards.

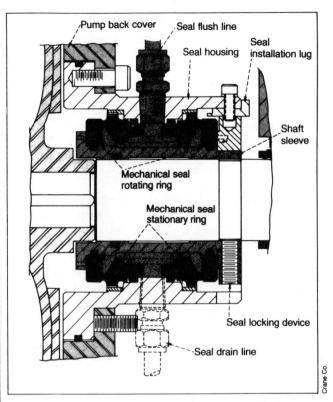

Figure 4 — Symmetrical packaged seal has stationary double seal

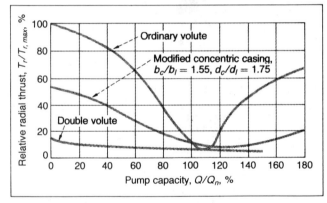

Figure 3 — Radial thrust in pumps varies with casing design

fits into the cover-plate recess intended for the symmetrical packaged seal. If the stuffing box is constructed conventionally — i.e., integrally cast with the pump cover — its dimensions must exceed those of ANSI pumps to accommodate the newer seal types and configurations.

To thus upgrade the stuffing box, a principal feature of the ANSI pump must be surrendered: its standardized dimensional envelope. Because a genuine upgrading of ANSI pumps is not possible with adherence to the dimensional limitations of ANSI Standard B73.1, it would not be proper to label a pump that incorporates the suggested stuffing-box revision as an "upgraded ANSI design." Therefore, the designation "upgraded medium duty (UMD)" is proposed for such a pump.

The merits of design optimization far outweigh the advantages of standardized external dimensions. A reliable pump operated and maintained judiciously is not likely to need to be replaced with another manufacturer's during the life of a plant. Should this become necessary, however, the baseplate and piping modifications necessary to accommodate a UMD pump could generally be accomplished without major difficulty.

The opportunity to improve hydraulic performance represents another reason for giving up the ANSI B73.1 dimensional envelope. There is ample evidence to support the contention that, without the ANSI B73.1 dimensional limits, the efficiency and net positive suction head (NPSH) characteristics of chemical process pumps can often be improved.

Upgrading bearing and lubrication systems

Unlike API pump bearings, which often are specified for a B-10 life of 40,000 hours, ANSI pump bearings are selected on the basis of a 24,000-hour life expectancy. Nominally, this means that 90% of the ANSI pump bearings should still be serviceable after approximately three years of continuous operation. However, Exxon's failure statistics indicate that conventionally lubricated ANSI pump bearings do not even approach this longevity. Lack of lubrication, wrong lubricants, water and dirt in the oil, and oil-ring debris in the oil sump all cause bearing life expectancies to be substantially less.

Problems caused by dirt and water have been substantially improved by oil mist lubrication. However, regardless of how the lubricant is applied, pump reliability can be further improved by specifying the following two types of bearings:
- Deep-groove Conrad-type radial bearings having loose fits (Class 3), and machined bronze cages. This type is more tolerant of off-design pump operation than the standard bearings customarily stocked by local suppliers.

• A pair of angular-contact (40-deg) thrust bearings, mounted back-to-back, having a light preload and Class 3 fit. With few exceptions, this type of bearing will not only prevent axial rotor shuttling and ball skidding, but also protect against mechanical seal distress.

The overwhelming majority of ANSI pumps, and even many API pumps, do not comply with the foregoing requirements. Instead, they frequently are equipped with the arrangement shown in Fig. 5, which consists of a double-row thrust bearing inserted into a cast-iron bearing housing, with the location of the thrust bearing determined by a snap ring.

Double-row thrust bearings in centrifugal pumps have not, generally, performed as well as two single-row thrust bearings mounted back-to-back [4]. The reason for this is probably not a fundamental weakness in the design of double-row bearings but rather the need for loose internal fitting (Class 3) of double-row bearings (as pointed out in many pump repair manuals), a requirement that is sometimes disregarded or overlooked by maintenance personnel.

Because of the tight internal fit, the bearing is extremely sensitive to skewing of races. Skewing is relatively frequent in small centrifugal pumps and in similar equipment liable to shaft misalignment and related abnormalities. Unless the amount of lube oil reaching a skewed race is increased beyond normal delivery limits, the bearing will overheat and fail. Locating the bearing by means of a snap ring makes matters worse.

Less than 10% of all ball bearings run long enough to succumb to normal fatigue failure, according to the Barden Corp. [5]. Most bearings fail at an early age because of static overload, wear, corrosion, lubricant failure, contamination, or overheating. Skidding (the inability of a rolling element to stay in rolling contact at all times), another frequent cause of bearing problems, can be eliminated by preloading a bearing.

Actual operations have shown that better bearing specification practices will avert the majority of static overload problems. Other problems caused by wear, corrosion, and lubricant failure, contamination and overheating can be prevented by the proper selection, application and preservation of lubricants. Oil viscosity and moisture contamination are primary concerns, and higher-viscosity lubricants are generally preferred [6]. The detrimental effects of moisture contamination are indicated in Table I [7].

Magnetic shaft seals in lubrication systems

Most pump shaft seals are generally inadequate. ANSI pumps are usually furnished with elastomeric lip seals. When these seals are in good condition, they contact the shaft and contribute to friction drag and temperature rise in the bearing area. After 2,000 to 3,000 operating hours, they are generally worn to the point at which they no longer present an effective barrier against contaminant intrusion. Percent failure vs. hours to leakage for two types of lip seals is shown in Fig. 6 [8].

A solution to the problem of sealing pump-bearing housings can be found in aircraft and aerospace hydraulic pumps, which make extensive use of magnetic face seals. This simple seal consists of two basic components (Fig. 7): (1) a magne-

Table I — Water in lubricating oil reduces rolling contact life

Base oil description	Water content of wet oil, %	Fatigue-life reduction, %	Test equipment and Hertzian stress
Base mineral oil dried over sodium	0.002	48	Rolling 4-ball bearing, 8.60 GPa (1.25 × 10^4 psi)
	0.014	54	
	3.0	78	
	6.0	83	

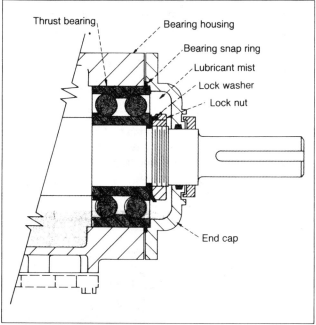

Figure 5 — Double-row thrust bearing is installed in most ANSI pumps

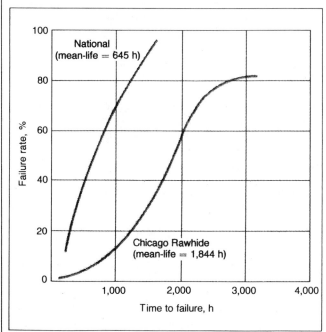

Figure 6 — Data from Ref. 8 show that elastomeric lip seals for pump shafts have a relatively short life.

tized ring, having an optically flat sealing surface, that is fixed in a stationary manner to the housing and sealed to the housing by means of a secondary O-ring, and (2) a rotating ring, having a sealing surface, that is coupled to the shaft for rotation and sealed to the shaft with an O-ring.

The rotating ring, which is fabricated from a ferromagnetic stainless steel, can be moved along the shaft. When no fluid pressure exists, the sealing surfaces are held together by the magnetic force, which is reliable and uniform, creating a positive seal with minimum friction between the sealing faces and ensuring the proper alignment of the surfaces through the equal distribution of pressure.

Magnetic seals are very reliable. Many have operated continuously for 40,000 hours without repair or adjustment under conditions considerably more severe than those to which chemical process pumps are typically exposed.

Finally, if sealed-bearing housings must accommodate changes in internal pressure and vapor volume, an expansion chamber similar to that shown in Fig. 8 can be used. This small device — which incorporates an elastomeric dia-

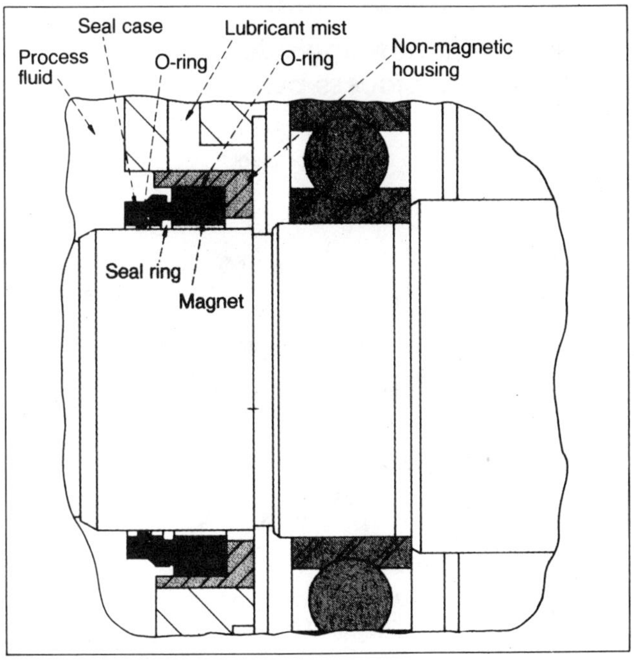

Figure 7 — Magnetic seals have proven to be extremely reliable

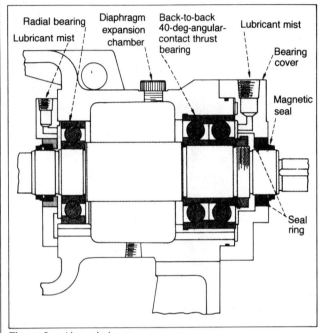

Figure 9 — Upgraded pump incorporates these bearing-housing features

phragm and constitutes a completely enclosed system — is screwed into the housing vent opening. It accommodates the expansion and contraction of vapors in the bearing housing without permitting moisture and other contaminants to enter. Carefully selected from a variety of plain and fabric-supported elastomers, the diaphragm will not fail prematurely in harsh chemical environments [9].

The upgraded medium-duty bearing housing shown in Fig. 9 incorporates the various bearing-related features that have been discussed: (1) a deep-groove Conrad-type, loose-fitting radial bearing; (2) a duplex angular contact (40-deg), lightly preloaded back-to-back-mounted thrust bearing; (3) a vent port that remains plugged for dry-sump oil-mist-lubricated bearings and that can be fitted with an expansion chamber if conventionally lubricated; (4) a magnetic seal; and (5) a bearing housing end cover made to serve as a directed oil-mist fitting.

Cast steel for pressure-containing parts

For safety reasons, the pressure containment parts of medium-duty pumps — the intermediate (pump-to-power frame) bracket and the power frame — should be constructed of a non-frangible material, such as W.C.B. cast steel (a weldable and castable low-carbon steel).

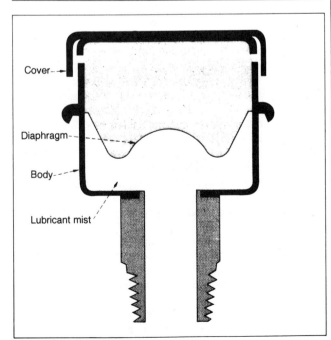

Figure 8 — Expansion chamber accommodates changes in vapor volume

A pump with cast iron parts, as shown in Fig. 10, was shut down and left full of liquid. Afterward, it was inadvertently started against a closed discharge block valve. As expected, the liquid vaporized. Unexpectedly, the casing became overpressured and blew apart [10].

Failure analysis quickly disclosed that, in this particular pump model, an intermediate bracket of cast iron was supposed to hold the cover against the pump casing. The casing blew apart simply because the intermediate bracket was not designed to contain pressure.

This design also invites uneven application of torque to the containment bolts. This could cause the unsupported bolt-hole flange incorporated in the cast-iron distance piece to break. The gap F between the stuffing box and the frame adapter/intermediate shown in the Fig. 10 pump does not represent sound construction.

A similar pump incorporating the design features that will preclude the foregoing problems is illustrated in Fig. 11. Proper design and construction of W.C.B. cast steel as the minimum-grade material can completely eliminate these

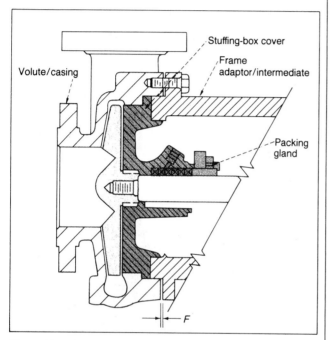

Figure 10 — Gap F invites rupture and uneven torque on bolts

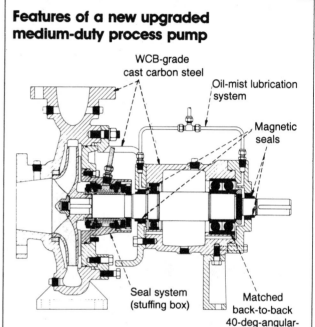

Features of a new upgraded medium-duty process pump

The pump shown above can be characterized as a foot-mounted chemical process pump designed for operation at casing pressures up to 300 psig at temperatures to 350°F. It incorporates the following standard features:

1. Improved shaft design, with shaft deflection not exceeding 0.001 inch when operating at any capacity between 0.25 and 1.25 best operating points.
2. Stuffing box accommodating all currently available seals and seal accessories.
3. Magnetic face seals in place of lip seals, for operations in hostile environments.
4. Matched back-to-back 40-deg angular contact bearings for higher thrust-carrying capacity.
5. Intermediate bracket and power frame of W.C.B. steel.
6. Baseplate that supports pump and motor so that pump operates without baseplate deflection, excessive vibration, or resonance.
7. Nozzle-loading capability exceeding that of ANSI B73.1 pumps.

Offered as routinely available options: power-frame bearing covers that accommodate a mist-lubrication system, and all API seal piping arrangements.

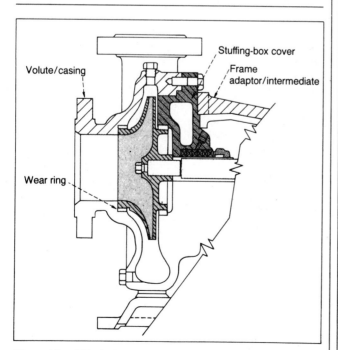

Figure 11 — This design corrects the shortcomings of Fig. 10 pump

risks. The additional cost is relatively low compared with the safety benefit realized.

Upgrading the baseplate is not difficult

Cost-cutting in design and manufacturing has resulted in weak baseplates being furnished with many ANSI pumps. Baseplate problems can be avoided by requiring the full disclosure of baseplate design details and by comparing material thickness, stiffener sizing and locations, grout-hole dimensions, and paint and primer protection provided by the baseplate designs of the pumps submitted by vendors. It is possible to obtain a baseplate that can be solidly grouted to the foundation and that will neither warp nor resonate while in service.

Table II — Upgrading extends component mean time between failures

Pump components	Average MTBF observed for ANSI pump, yr	Expected MTBF for upgraded pump, yr
Ball bearings, L_1	3	5
Shaft, L_2	10	15
Mechanical seals, L_3	1.2	2.5
Coupling, L_4	4	7

Table III — Lengthening component life increases pump onstream time

	Component MTBF, yr				Pump MTBF, yr
	L_1	L_2	L_3	L_4	
Best available ANSI pump	3	10	1.2	4	1.07
Upgraded medium-duty process pump	5	15	2.5	7	2.11

Mean-time-between-failure calculations

The problem of mechanical seal life was investigated by making an assessment of probable failure avoidance that would result if shaft deflections could be reduced. It was decided that a maximum 0.001-inch deflection of the seal face would probably increase seal life by 10%. It was similarly judged that a sizable increase in seal housing dimensions to allow the installation of the newest seal configurations would more than double the mean time between failures of seals.

By means of such analysis, all the components under consideration for upgrading were examined. New cost data and life estimates were collected, and the latter were used in mean-time-between-failure calculations.

Mean time between failures (MTBF) is calculated as follows:

$$\frac{1}{MTBF} = \left[\left(\frac{1}{L_1}\right)^2 + \left(\frac{1}{L_2}\right)^2 + \left(\frac{1}{L_3}\right)^2 + \left(\frac{1}{L_4}\right)^2\right]^{0.5}$$

Here, L = life, in years, of the component subject to failure.

Using the data in Table II, which is representative of that collected by Exxon, mean times between failures were calculated. These results, and estimated values for an upgraded medium-duty pump, are presented in Table III.

As has been noted, a pump failure today, based on actual reports, costs $5,000 or more. This includes costs for material, parts, labor and overhead. If the MTBF for a particular pump were 12 months, and it were extended to 24 months, this would, of course, result in a cost avoidance of $2,500/yr, which is greater than the premium one would pay for the upgraded medium-duty pump.

In addition, the probability of reduced power cost would, in most cases, further improve the payback. The elimination of the ANSI dimensional envelope would make possible improvements of 10% or more in the operating efficiency of many hydraulically redesigned pumps. It was calculated that in one Gulf Coast plant these seemingly small efficiency gains could result in power-cost savings of several hundred thousand dollars per year.

Thus, the primary advantages of the upgraded medium-duty pump are longer online time, higher operating efficiency and lower operating and maintenance costs.

References

1. Cherry, R. C., "A Review of Design Analysis Methods for a Horizontal End Suction Centrifugal Pump," Proceedings of the Eighth Turbomachinery Symposium, Texas A&M University, Houston, Tex., December 1979, pp. 21-25.
2. Bloch, H. P., "A User's Views of Fluid Sealing Economics," Forty-fifth Annual Meeting of the Fluid Sealing Assn., Sun Valley, Idaho, October 1978.
3. Netzel, J. P., "Symmetrical Seal Design: A Sealing Concept for Today," Proceedings of the First International Pump Symposium, Texas A&M University, Houston, Tex., May 1984, pp. 109-112.
4. James, R., Pump Maintenance, *Chem. Eng. Prog.*, February 1976, pp. 35-40.
5. Mackenzie, K. D., "Why Ball Bearings Fail," Product Data Bulletin No. 5, The Barden Corp., Danbury, Conn.
6. Bloch, H. P., "Practical Machinery Management for Process Plants, Vol. 1: Improving Machinery Reliability," Gulf Publishing Co., Houston, 1982.
7. Armstrong, E. L., Murphy, W. R., and Wooding, P. S., Evaluation of Water-Accelerated Bearing Fatigue in Oil-Lubricated Ball Bearings, *Lubrication Engineering*, Vol. 34, No. 1, 1977, pp. 15-21.
8. Horve, L. A., "CR Waveseal — A Detailed Synopsis of Five Comparative Life Tests," CR Industries, Elgin, Ill., 1977.
9. Nagler, B., Breathing: Dangerous to Gear Case Health, *Power Transmission*, January 1981.
10. Bloch, H. P., "Mechanical Reliability Review of Centrifugal Pumps for Petrochemical Services," ASME Failure Prevention and Reliability Conference, Hartford, Conn., September 1981.
11. Grunberg, L., and Scott, D., The Effect of Additives on the Water-Induced Pitting of Ball Bearings, *J. of Institute of Petroleum*, Vol. 46, 1960, pp. 259-266.

The authors

Heinz P. Bloch is a senior engineering associate with Exxon Chemical Co. (Central Engineering Div., Baytown, TX 77522; tel: 713-425-2257). Many of his assignments have centered on machinery reliability assurance, condition monitoring, and maintenance cost reduction, and he has written many papers and published articles, as well as several books, on the subject. He holds B.S. and M.S. degrees in mechanical engineering from the New Jersey Institute of Technology.

Donald A. Johnson is vice-president of sales, marketing and engineering for Carver Pump Co. (2415 Park Ave., P.O. Box 389, Muscatine, IA 52761; tel: 319-263-3410). Previously, he held executive positions with Fairbanks, Morse and Co., Johnston Pump Co., and Studebaker, Worthington Corp. He graduated from the University of Minnesota with a B.S. degree in aeronautical engineering.

Standard pumps are *not* obsolete!

Recently, there has been a move to abandon the ANSI chemical-pump standards because of claimed inadequacies of present standard pumps. But the standard pump has too many advantages to be lightly discarded.

John A. Reynolds
Union Carbide Corp.

Recently, ANSI (American National Standards Institute) B73 centrifugal pumps have been getting a bad press [1]. Detractors say that ANSI pumps require a lot of maintenance and that several design changes are required to improve mean-time-between-failures from an industry average of nearly six months to twenty-five months or more. (See the preceding article.)

The most radical change suggested is to depart from the ANSI standard dimensions. Other changes include:

1. Increasing the shaft diameter to limit shaft deflection at the face of the stuffing box to 0.001 in. when operating between 25% and 125% of the best-efficiency flowrate.

2. Always using a pair of back-to-back 40-deg-angular-contact thrust bearings.

3. Using separate bolt-on packaged-seal assemblies to provide more room for complex seals.

4. Applying contact-type magnetic seals to protect the bearing housing. An expansion chamber mounted on the housing prevents in-breathing.

5. Altering the bearing frame adapter to provide better pressure containment of the casing.

6. Strengthening the pump baseplate and adequately grouting it in place.

API (American Petroleum Institute) refinery pumps are discussed as not being cost effective in medium-duty service.

All of these changes may be appealing to the ANSI pump user who has spent many hours in the field or maintenance shop trying to solve a tough pumping problem. But, note that all of these features can be obtained in pumps made under the present ANSI pump standard—the buyer need only include them in the purchase specifications. Changes to the B73 standard are not required to obtain them. Revision of the standard is required only when these "specials" become commonplace. In any event, before rushing out to order one of these new medium duty or so-called "upgraded" pumps, please read on.

Mean time between failures

Our company's first response to such bad news about the high failure rate of ANSI pumps was to collect failure statistics to see how Union Carbide's chemical pumps were faring. The more-mature plants have many pre-ANSI pumps, some less rugged than ANSI ones but which could not be eliminated from the statistics. A survey of 1985 maintenance data from two larger and mature chemical plants and a new, small chemical plant revealed the numbers that are in the table.

These figures were a pleasant reassurance. We had arrived at, or surpassed, the longevity goal set by designers of "upgraded" pumps without having to go on a spending spree—or having to obsolete our pumps for a new generation of models, or emptying our storerooms of valuable spare parts. Our engineering department could continue to use the standard outline dimensions and to forego the expensive piping and foundation revisions that were so common before the advent of the first "AVS" (American voluntary standard) pumps some 25 years ago.

Dimensional standards

Perhaps pump dimensional standards are not important to some users, but in a competitive industry, time is money, and if the equipment dimensions are known at project approval time, then engineering can forge ahead, bypassing the procurement cycle and shortening project time. Dimensional standards are a most important feature of the ANSI pump.

Besides these engineering advantages, when the pumps are operating and need to be removed for maintenance, replacements can be supplied by several manufacturers without requiring costly piping and foundation alterations. Indeed, the minimizing of plant downtime caused by adapting from one pump manufacturer's dimensions to another's was one of the major reasons for the chemical process industries' overwhelming acceptance of the standard pump.

Originally published May 12, 1986

STANDARD PUMPS ARE *NOT* OBSOLETE!

Table — 1985 survey of pump failures at three Union Carbide plants

Plant	Total pumps	1985 failures	Failures, %/yr	Meantime between failures, yr
Plant A	4,772	542	11.4	8.8
Plant B	4,595	1,018	22.2	4.5
Plant C	120	55	45.8	2.2

Pump size	T (Packing)	T (Mechanical seal)
AA - AB	5/16 in.	3/4 in.
A05 - A80	3/8 in.	7/8 in.
A90 - A120	7/16 in.	1 in.

Figure — Proposed ANSI/ASME B73. pump stuffing-box dimensions

Imagine the resulting chaos if the electrical manufacturers dropped the NEMA motor-frame dimensional standards!

Functional standards

To follow the NEMA analogy further, NEMA does not dictate internal design; neither does ANSI B73. The B73 standards specify 0.002-in. maximum shaft deflection at any point on the curve, with maximum impeller diameter (except for the very largest pump sizes). Thus, if the pump is operating anywhere in the better efficiency areas, the shaft deflection is going to be 0.001 in. or less. Similarly, a minimum bearing-life of two years under maximum load is specified. This means that our pumps are enjoying much longer bearing lives, as our statistics illustrate. None of our pumps have duplex 40-deg-angular-contact thrust bearings or bearing housings with magnetic contact-type seals. ANSI B73 says that the "Bearing housing shall be constructed to protect the bearings from water, dust, and other contaminants." The specific details of designing B73 pumps are left to the manufacturer's engineering judgment and the acceptance of the buyer—who is free to dictate preferences.

Rugged design does not always assure longer life. Many pumps are ruined by careless operators and maintenance personnel. The fact that Union Carbide's mechanical-seal mortality greatly exceeds bearing failure attests to this. Our two mature plants have been using lubricated-for-life bearings for several years to combat lack of proper lubrication in the plant. But mechanical seals cannot be lubed-for-life, and most, being single seals, have to survive in the harsh environment of dirty and aggressive process liquids. Their very lives depend upon an operator applying the proper flush and not running them dry, not feeding them doses of rust and pipe-scale, and not shocking them thermally or hydraulically. (For more on pump seals, see Part IV, page 105.)

ANSI B73 Standards committee

The ASME B73 committee is "balanced" as defined by ANSI's rules, with pump users, pump producers and general-interest groups, so that any one group cannot dominate. The consensus of these groups has prevented setting the requirements beyond a basic machine that can perform a fairly well defined job. Many users have supplemented the B73 standards for particular applications by specifying: oil-mist lubrication or special bearing-housing seals for humid or dirty atmospheres, stainless-steel bearing-housing and frame adapters for corrosive environments, foot-supported baseplates to permit greater piping flexibility and reduce installation costs, and extra-large stuffing boxes for special mechanical seals, just to name a few. These additions have never been a real problem for pump manufacturers.

In fact, as noted earlier, all the six changes listed in the second paragraph of this article as reasons for abandoning the ANSI Standard pump can be accommodated by the present B73 standard. The B73 committee tries to keep abreast of these "specials", so that when interest is sufficient, the standards can be modified to address them. Currently, larger stuffing-box-dimension standards have been developed [2] to accommodate the more-complex cartridge-type single, double and tandem seals, and will be included in the next revision of B73 (see figure). A separate bolt-on stuffing box was considered but rejected because it added another rather large joint as a possible leak source. Some users are already specifying the larger stuffing boxes. Baseplate rigidity was addressed in the 1984 edition of Standard B73.1. An optional ductile-iron or carbon-steel frame adapter is currently being considered. All this illustrates that B73 standards are functional, flexible to meet special situations, and are evolving to recognize new technology.

Let us not abandon the B73 standards and the thousands of ANSI pumps that are faithfully serving us.

References

1. Bloch, H. P., and Johnson, D. A., Downtime Prompts Upgrading of Centrifugal Pumps, *Chem. Eng.*, Nov. 25, 1985, pp. 35–41.
2. ASME Codes & Standards [proposed change in stuffing-box dimensions for ANSI Standard B73], *Mech. Eng.*, June 1985, pp. 93, 95.

The author

John A. Reynolds is a senior staff engineer in the Machinery Engineering Technology Group at Union Carbide Corp.'s Chemical and Plastics Div. Technical Center, P.O. Box 8361, South Charleston, WV 25303. He has provided both primary procurement and technical support on pumps and pumping equipment for over 20 years for the Chemicals and Plastics Div., and is responsible for the application and specification of pumping equipment and mechanical seals used on major projects, and for solving equipment problems. His articles have dealt with mechanical evaluation of centrifugal pumps, pump installation and maintenance, pump energy conservation, and mechanical-seal buffer-liquid systems. He holds a B.S.M.E. from West Virginia University, is chairman of the ASME B73 Pump Standards Committee and is a member of ASME's Board on Performance Test Codes.

Head-vs.-capacity characteristics of centrifugal pumps

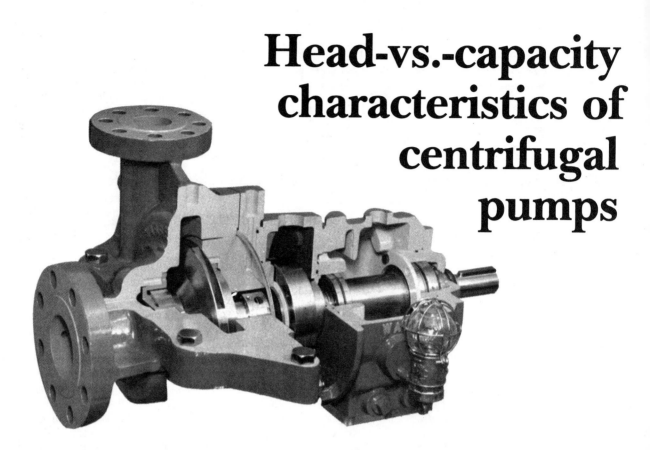

Engineering standards limit the use of pumps having head-vs.-capacity curves that do not rise constantly to shutoff. But in many cases such pumps are a good choice because of their high efficiency.

James J. Paugh, Warren Pumps Div., Houdaille Industries, Inc.

☐ Many pump standards such as API 610 require that centrifugal pumps operated in parallel have characteristic head-vs.-capacity curves that rise constantly to shutoff, as in Fig. 1a. However, high-head, low-capacity, single-stage pumps as used in the chemical process industries often have curves that do not do so: Rather, there is a peak head at some capacity, and from that point the head *falls* constantly to shutoff. Such a pump curve, shown in Fig. 1b, is known as a "drooping" curve.

This article explains why some centrifugal pumps have drooping head-vs.-capacity curves, and shows when and how such pumps can cause problems. The point, of course, is to prevent the problems.

As it turns out, the significance of a drooping curve is this: Two identical pumps having drooping head-vs.-capacity curves should not be operated in parallel under conditions that require wide ranges of capacity that approach zero. Such conditions are common in boiler-feedwater and fire-main systems. In chemical-process service, though, it is rare to find two pumps operating in parallel. They are often *piped* in parallel, but typically only one is operated at a time, and the other is a spare.

Originally published October 15, 1984

Thus, drooping-curve pumps can often be used in chemical-process applications. What is their advantage? Either greater efficiency or lower cost. Compared with a pump having a constantly-rising head-vs.-capacity curve, a pump of the same size but having a drooping curve will be more efficient. And, one of a smaller size can deliver the same capacity at the same head, with equal efficiency.

Head vs. capacity

Let us now look at head-vs.-capacity curves, and at why certain pumps do or do not have drooping curves. The head generated by a centrifugal pump can be expressed in the form:

$$h = AN^2 + BNQ + CQ^2$$

where h is head (ft), N is rotational speed (rpm), Q is flowrate (gpm), and A, B and C are constants for a given pump and impeller. For the typical case of a pump operating at constant speed, this head-vs.-capacity equation can be rewritten as:

$$h = a + bQ + cQ^2$$

The constants a, b and c are for a given speed.

HEAD-VS.-CAPACITY CHARACTERISTICS OF CENTRIFUGAL PUMPS

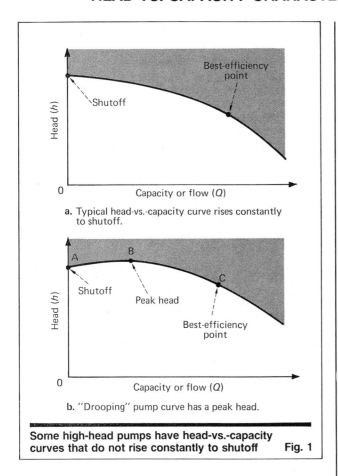

Some high-head pumps have head-vs.-capacity curves that do not rise constantly to shutoff — Fig. 1

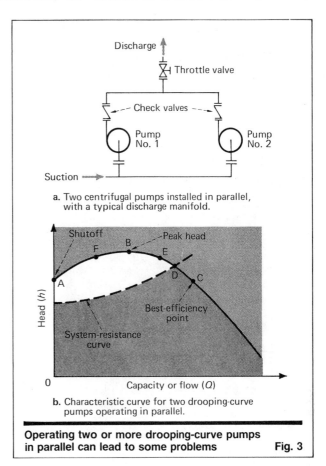

Operating two or more drooping-curve pumps in parallel can lead to some problems — Fig. 3

This equation describes a parabola having its axis parallel to the h axis, as in Fig. 1, and its apex at $Q = -b/2c$. If this apex is far enough to the right of shutoff, then a head-vs.-capacity test will show a detectable droop: The shutoff head (a) will be detectably lower than the peak head.

Whether a pump has a drooping curve or not is decided by the designer: Pumps are designed to be run at their best-efficiency point (see Fig. 1), and where the designer puts this point determines where the peak head falls.

It turns out that high-head, low-capacity pumps designed for optimum efficiency will tend to have drooping head-vs.-capacity curves. In pump terminology, pumps whose specific speed ($N_s = N(Q^{1/2}/h^{3/4})$) is less than 1,000 will tend to have drooping head-vs.-capacity curves if they have an optimum number of vanes (7 or 8) and an optimum discharge angle (23 – 27 deg).

A pump design can be modified to avoid drooping, but there is a penalty for doing so. Reducing the vane discharge angle (to, say, 18 deg) does away with drooping, as shown in Fig. 2, but it also creates higher fluid velocities in the impeller passages and thus reduces head and efficiency. To get the head back up requires a larger impeller.

So the drooping-curve pump has an advantage in either efficiency or capital cost. Considering today's high power costs and interest rates, it makes sense to use the drooping-curve pump. But when is it possible to do so?

Single and parallel pumps

A pump system should not have surges, i.e., undesirable swings in head, capacity and power. But surging can occur in a pumping system when three conditions are present. It rarely happens in a single-pump system, but is more common in systems having two or more pumps operating in parallel:

The mass of liquid must be free to oscillate. This condition exists when the mass of water (or other liquid) is suspended between two free surfaces, i.e., when the suction is taken from a vessel containing a free surface, and the discharge is to another such vessel. This is the case in a boiler feedpump system: suction from a feedwater heat-

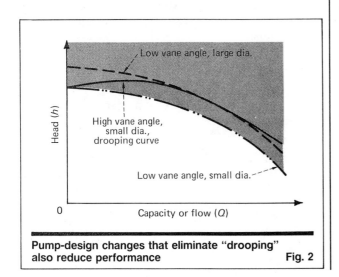

Pump-design changes that eliminate "drooping" also reduce performance — Fig. 2

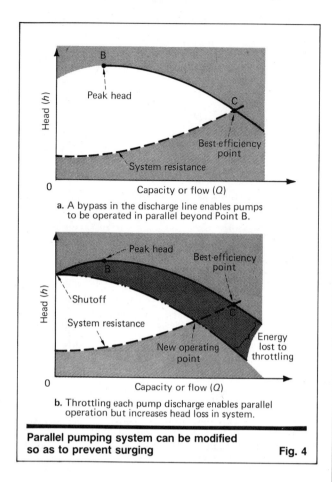

a. A bypass in the discharge line enables pumps to be operated in parallel beyond Point B.

b. Throttling each pump discharge enables parallel operation but increases head loss in system.

Parallel pumping system can be modified so as to prevent surging — Fig. 4

er, discharge to a header. Also in a condensate system: suction from a condenser, discharge to a deaerating feedwater heater.

Some part of the system must be able to store and give back pressure energy. In a boiler-feedpump system, the steam cushion in the boiler serves this purpose. So does a static water column, or entrained gas in a long piping system, when those are present.

Some part of the system must provide impulses to start the swing. Usually, this means another pump, operating in parallel. The impulse can come from startup, shutdown, or discharge throttling. [However, single-pump systems (e.g., for boiler feedwater) can swing from cavitating to noncavitating operation. This creates a "chugging" noise, and is accompanied by wide variations in flow and head. Of course, this can occur whether a pump has a drooping curve or not, but it is more violent in the case of a drooping-curve pump operated at or near shutoff.]

It is unusual for all three of these conditions to be present, except when pumps are operating in parallel. Let us consider what happens in that case.

Operating pumps in parallel

Fig. 3a shows a typical setup for two centrifugal pumps operating in parallel. Suppose that the pumps are identical, each having a drooping head-vs.-capacity curve; their combined curve (adding their capacities) is shown in Fig. 3b. Here are three problems that can occur:

First, suppose one pump is operating at a head (say, D) greater than the head at shutoff (A). Then the other pump cannot be put on line, as its head at zero capacity will not be enough to open the check valve against backpressure.

Now, suppose both pumps are operating at Point C, and flow demand is reduced by partly closing the throttle valve to Point E. In this case, one pump or both may operate at E, but one or both may also operate at F, where the head is the same but the flow is lower. In this case, the pumps will not be sharing the load equally, and flow and pressure may start fluctuating.

What happens if the throttle valve is now opened, in response to a demand for greater flow? A pump at Point E will increase its flow, but one at Point F will reduce its flow and may cease delivering entirely. This will cause pressure fluctuations, the impulse necessary to start a surging. If surging begins, it will stop only if the pump or pumps at F are shut down.

Surging in such a system can be prevented, though.

One way is to install in the discharge line a bypass that will pass all flow to the left of (i.e., less than) that at Point B, in Fig. 4a. Now, the system will operate only to the right of Point B (i.e., at greater flow and lower head), and so it will not start surging if head increases and flow decreases. The bypass will also protect the pump from overheating at low flowrates, when there might otherwise not be enough liquid to carry heat away.

Another way is to put a throttle valve or orifice on each of the discharge lines, either in place of or in addition to the single valve on the combined discharge. This changes the effective pumping curve to a constantly rising one, as shown in Fig. 4b. However, this will take more power, due to the presssure drop added by the valves or orifices.

As for operating at low capacity—to the left of the peak head—it is not advisable to operate drooping-curve pumps in parallel in this range.

Conclusions

Summing up, there are some problems associated with using high-head, low-capacity pumps that have drooping head-vs.-capacity curves. However, the situations that cause problems are rare in chemical process applications—pumps are not often operated in parallel, even if they are installed in parallel. And the problems may be prevented in most cases. Thus, since drooping-curve pumps can provide greater efficiency, or save on capital, it is worthwhile to consider whether they can be used in a given situation.

The author

James J. Paugh is Vice-President of Engineering for the Warren Pumps Div. of Houdaille Industries, Inc., Bridges Ave., Warren, MA 01083; tel. (413) 436-7711. He has had over 25 years of experience in the pump industry, and is a registered professional engineer in Massachusetts and New Jersey. Mr. Paugh earned his B.S. degree in mechanical engineering at Bucknell University. He is his company's principal technical representative to the Hydraulic Institute, and he belongs to ASME, TAPPI and ASTM.

Multistage centrifugal pumps

Centrifugal pump bodies can be combined in one casing to gain head at constant flowrate. This style, particularly the "donut" type, has several useful applications.

S. Yedidiah, McGraw-Edison Co., Worthington Div.

☐ Single centrifugal pumps, despite their universal popularity in a wide range of applications, can sometimes be inefficient when pumping a small flow against a high head. In such cases, multistage centrifugal pumps may prove useful. There are two main types—the split-casing type and the "donut" type. This article will concentrate on the advantages of the donut type.

Centrifugal pumps

Of all the devices used for transporting liquids, none seems to have gained greater popularity and acceptance than the centrifugal pump. In spite of this widespread use, there are still many applications for which the centrifugal pump is at a disadvantage in relation to other liquid-moving devices.

One such application arises when a pump has to deliver a relatively small amount of liquid against a very high head. In such cases, the efficiency of a centrifugal pump becomes prohibitively low, and it becomes necessary to look for another type of pump.

Generally, the maximum efficiency attainable with a given centrifugal pump depends on both its size and the relationship between head and flowrate at a given speed. This relationship is usually expressed in terms of the specific speed, N_s, which is given by:

$$N_s = NQ^{0.5}H^{-0.75} \qquad (1)$$

where N is the operating speed of the pump in rpm, Q is the flowrate in gpm, and H is the head in ft.

Fig. 1 represents the efficiencies attainable with centrifugal pumps, as a function of their size (flowrate) and their specific speed.

As an example, let us consider two pumps, each designed for 300 gpm, but different heads. Let Pump 1 be designed to develop a head of 230 ft at 3,560 rpm, and Pump 2 to deliver the same flowrate at the same speed, against a head of 1,150 ft.

Using Eq. (1), we find that the specific speed of Pump 1 is 1,044, while that of Pump 2 is only 312.

From Fig. 1, we find that Pump 1 can be expected to attain an efficiency of 72%, while the efficiency of Pump 2 will be only 44%. Thus Pump 2, operating against the higher head, pumps the same quantity of liquid at considerably higher cost. Assuming that the two pumps operate 3,000 hours per year, and that the cost of electricity is 5 cents per kWh, the cost difference is more than $8,000 per year.

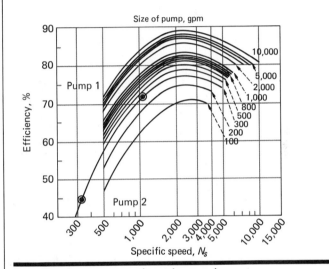

Centrifugal pumps pass through a maximum efficiency at a particular specific speed   Fig. 1

This expense could be avoided by replacing Pump 2 with five pumps such as Pump 1, operating in series. Each pump would consecutively add 230 ft to the head, resulting in a total head of 1,150 feet. This solution, however, is impractical because of the prohibitively high initial investment and very high maintainance costs.

Multistage centrifugal pumps

This conundrum led to the development of a special class of pumps, generally known as multistage pumps. A multistage pump is essentially a combination of several pumps, operating in series, made into one integral unit. In such a pump, all impellers are mounted on one common shaft, with one casing.

There are two basic types of multistage pumps:

1. The multistage split-casing pump, in which all casings are combined into one complicated casting. This casting is made of two parts, split in a horizontal plane. The lower part contains the pipe connections and the support, and the upper part, often called the cover, is held firmly against the lower by means of bolts. The main advantage of such a pump is that it allows access to the inner parts by simply opening up the cover, without disturbing the alignment between the pump and the

CENTRIFUGAL PUMPS

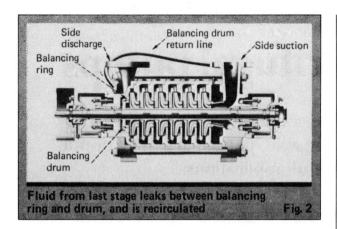

Fluid from last stage leaks between balancing ring and drum, and is recirculated — Fig. 2

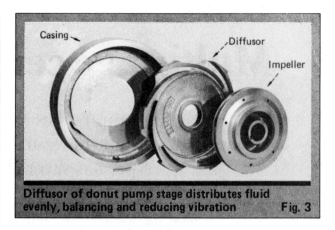

Diffusor of donut pump stage distributes fluid evenly, balancing and reducing vibration — Fig. 3

driver, and without the need to disconnect the piping.

2. The multistage donut pump (Fig. 2), in which each casing forms a donut-shaped unit, and all individual "donuts" of a complete pump are held together by means of long bolts.

In the U.S., the multistage split-casing pump is very popular for relatively large flowrates. In pumps designed for lower flowrates, however, the losses incurred in the long and narrow passages of the split-casing pumps significantly offset the efficiency gained through multistaging. In such cases, it becomes more economical to use donut pumps.

Donut pumps

Although donut pumps lack the convenience of easy access to the internal parts without removal of the pump from the system, they do possess a number of other advantages that compensate for this deficiency.

A donut pump consists of a number of identical elements, called stages. Each stage consists of a casing (C in Fig. 3), a diffusor backflow combination (D), and an impeller (I). All impellers are mounted on a common shaft and located within their respective casings, forming one integral unit (Fig. 2, 4).

The flowrate delivered by such a multistage pump is equal to the flowrate of one individual element, but the head is equal to the head of one stage multiplied by the number of stages.

The donut pump has a wide range of applications, due to the freedom of assembling it with any required number of stages—as long as the maximum mechanical limitations are not exceeded. Consequently, the same parts may be used for making up pumps to operate against different heads.

When the total required head is not an integral multiple of the head produced by a single stage, adjustments can be made by reducing the outside diameter of one or more impellers.

Operation at shutoff

Because the diffusor of a donut pump is provided with vanes evenly distributed around the periphery of the

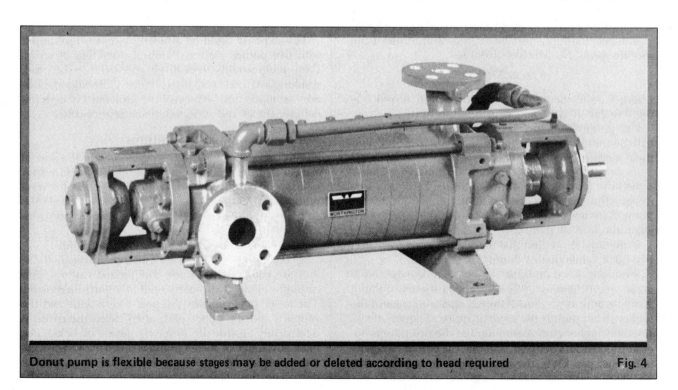

Donut pump is flexible because stages may be added or deleted according to head required — Fig. 4

impeller, the radial loads on the rotating element are reduced to a minimum. In a well-designed unit, this fact allows the pump to operate quietly and smoothly from its maximum flowrate down to nearly complete shutoff.

No pump should be allowed to operate with the discharge shut off for a prolonged time, as the heat generated due to turbulence and friction may raise the temperature of the contained liquid above its boiling point, and this may cause damage to the pump. In pumping water, it is generally recommended that at least 30 gpm should be allowed to flow through the pump for each 100 hp consumed, in order to prevent overheating. Some donut pumps have a built-in feature that automatically takes care of this, as follows:

In order to balance the (often huge) axial load on the impellers, donut pumps are usually provided with a hydraulic arrangement that consists of a balancing drum and a balancing ring (Fig. 4). When the pump is in operation, a certain amount of liquid flows through the gap between the balancing drum and ring, and this generates a pressure differential on both sides of the drum. This differential, in turn, balances the axial forces on the impellers, which have a low pressure at intake and a high pressure at discharge.

Under ordinary operating conditions, the liquid leaking through this balancing device is returned to the suction inlet of the pump. However, when a pump is destined to operate under conditions that mandate frequent and prolonged operation against a shut-off valve, it is often advantageous to recirculate the liquid leaking through the balancing device back via a suction tank, where it can cool off and prevent overheating.

If a pump that has no such central balancing device has to be operated frequently against a shut-off valve, it should be provided with a bypass that will allow some of the pumped liquid to escape continuously, although this results in a significant waste of energy.

Versatility

Another feature that adds to the versatility of a donut pump is the freedom of changing the position of the discharge and the suction nozzles (Fig. 5). Some donut pumps can be fitted with an end-suction casing, which not only adds to the flexibility of the optional nozzle positions, but also achieves better suction performance.

Moreover, the position of each nozzle can be altered independently from the other. This makes it possible to simplify the piping layouts to and from the pump and, consequently, to save space and labor.

Very often donut pumps can solve problems that otherwise would require replacing existing pumping units, or using multiple pumps.

In one case, for example, a pump had to operate for a limited period of time against a low head. However, it was destined to operate against a final head that was so high that it would have been uneconomical to use any other than a multistage pump.

The problem was solved by supplying a donut pump designed to operate against the total final head, but with several of its stages replaced by cylindrical spacers, without any impellers. This enabled the pump to operate efficiently against the temporary low head. Later, the cylindrical spacers were replaced by actual stages, which enabled the pump to operate satisfactorily against the full final head.

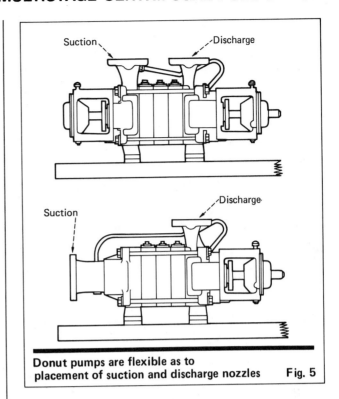

Donut pumps are flexible as to placement of suction and discharge nozzles. Fig. 5

The donut pump has also proven very useful in power-recovery applications, where it is used in reverse as a turbine.

The predominant factor that determines the net positive suction head (NPSH) requirements of a pump at a given operating speed is the flowrate for which it has been designed.

For end-suction pumps operating at 1,770 rpm this relationship can be expressed by:

$$NPSH = 0.6 Q^{0.424} \qquad (2)$$

Regular side-suction pumps require an NPSH about 50–60% higher, while end-suction donut pumps with inducers operate at about 80% of this NPSH.

References

1. Yedidiah, S., "Centrifugal Pump Problems, Causes and Cures," PennWell Book Pub. Co., Tulsa, Okla., 1980.

The author

Samuel Yedidiah is a hydraulic specialist at McGraw-Edison, Worthington Div., 14 Fourth Avenue, East Orange, NJ 07017; tel. (201) 484-2600, where his duties include serving all elements of the corporation as an expert in areas of hydraulic design. Active as a pump specialist since 1938, he has experience with all phases of pumps, from research and development, through testing, production, planning, management, and troubleshooting. A member of ASME, he has published many technical papers and holds a number of patents.

TIME TAKES ITS TOLL ON CENTRIFUGAL PUMPS

Just as no person can escape the effects of aging, neither can a pump. Here's how the latter's performance can change over time.

S. Yedidiah, Consultant

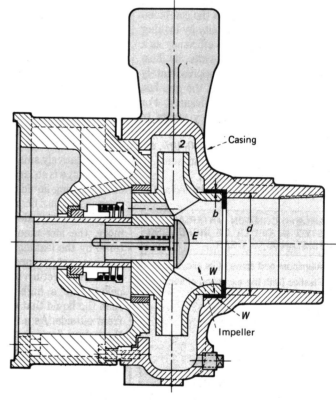

FIGURE 1. Cross-sectional view of an end-suction centrifugal pump

When selecting a centrifugal pump, one must remember that the unit's performance will rarely remain constant. Rather, it will be continuously changing.

Such change can be due to external causes, such as a change in demand due to fluctuations in usage, or the accumulation of sediment in the pipelines. Or it can be a result of internal causes — such as an increase in the gap between the pump's wear surfaces, a change in internal-surface roughness, or changes in the size of a pump's flowpaths. This article focuses on the effects of these three internal factors.

While there is not much that one can do to prevent pump wear (aside from periodic flushing and selection of the appropriate materials), one must understand the reasons for changes in pump efficiency, and must know where to look and how to spot the effects of wear. So informed, the engineer can then have the appropriate worn parts replaced and the pump's normal operation restored.

Originally published June 20, 1988

Wear increases leakage

Consider the end-suction closed-impeller pump shown in Fig. 1. The liquid enters the impeller at Zone **E** and leaves at Zone **2**. But because the liquid pressure at the outlet is higher than at the inlet, there is a tendency for the liquid to flow from Zone **2** to Zone **E**. To minimize such backflow, both the impeller and the casing are provided with wear surfaces, labeled **W**, and the clearance, **b**, between these two wear-faces is kept as small as possible. The amount of leakage, Q_L, that flows between the wear-faces is given by:

$$Q_L = K\pi db(2gH)^{1/2} \quad (1)$$

where b is the width of the gap, d is the mean diameter of the gap, H is the total head developed by the pump, g is the acceleration due to gravity, and K is a coefficient whose magnitude depends on a number of design features of a given pump and varies from 0.35 to 0.65.

In a new pump, the clearance b is very small. But over time, this clearance increases due to wear. The rate at which this wear takes place depends on the pump's operating conditions and certain design features. Examples of such operating conditions include:

- Suspended solids within the

pumped liquid. The higher the concentration of solids, the higher the rate of wear. However, the wear rate also depends on the hardness, shape and size of the particles.

• Dissolved solids within the pumped liquid. During standstill, some of these solids may recrystallize due to changes in temperature or to evaporation. When the pump is restarted, the solids may cause serious wear to the gap between the wear surfaces. To prevent this, the pump should be drained and flushed with an appropriate solvent after each stoppage.

• Noise and vibrations. The most common causes of noise and vibration are misalignment between the pump and driver, a poor foundation, stresses imposed on the pump by the pipelines, and cavitation.

One design factor that affects the rate of wear is the shape of the gap. Two shapes are common, an L-shaped gap and a straight gap. Experience has shown that the L-shaped gap (shown in red in Fig. 1) is significantly more resistant to wear than a straight gap would be. In addition, resistance to wear increases with the length of the gap (l).

Another important design aspect is the choice of suitable materials. This is often a complex problem, which focuses on the chemical compatibility between the wetted pump parts and the fluid being pumped. Some manufacturers have developed certain proprietary materials that, they claim, are esspecially suitable for certain applications. In the absence of other data, the table can serve as a general guide to the wear resistance of different materials.

Leakage reduces performance

The manner in which an increase in leakage will affect pump performance depends on several factors, including the specific speed of the pump ($N_s = NQ^{0.5}/H^{0.75}$, where N is the actual speed) and the width of the gap, b. For small values of b, the head-capacity curve will change approximately as shown in Fig. 2. As the width of the gap increases due to wear, so does leakage flow. Thus, for a given head H_1, the pump will deliver a new flowrate, Q_2, which is the original flowrate, Q_1, minus the increase (q) in the leakage flow. Conversely, the head H_2 at flowrate Q_2 will be reduced to H_1.

This phenomenon can be better understood if one considers the combined effects on head and capacity by looking at the variation in pump efficiency as a function of sealing-gap width, as illustrated in Fig. 3. Up to approximately $b = 0.01$ in., the width of the gap has virtually no effect on efficiency. For larger clearances, efficiency starts to drop nearly linearly with b. Then, after an efficiency drop of about 30–40%, additional increases in b again seem to have no effect on efficiency.

This behavior can be explained as follows: For values of b less than about 0.01 in., the boundary layers on both sides of the gap usually fill the entire width of the passage. Therefore, changes in the gap width have a negligible effect on the total flowrate delivered by the pump.

For b greater than about 0.01 in., the flow typically increases almost linearly with the width of the gap. However, as long as the absolute magnitude of b remains relatively small, the resistance to that leakage is so large that all the head of the leaking liquid is lost, and efficiency drops almost linearly with b.

As the magnitude of b increases still more, the incremental losses encountered by the leaking liquid are getting progressively reduced. The leaking liquid starts to return to the eye of the impeller with a higher energy content than the liquid that enters the impeller from outside. As a result of mixing of these two flows, some of the energy lost due to leakage is being returned to the incoming liquid. This results in smaller losses in efficiency. At very large values of b, an equilibrium sets in, and the loss in efficiency becomes almost constant.

It should be noted that Fig. 3 is based on the author's own experiences, and it should serve only as a general illustration of the effect of gap width on performance. In actual operation, such an effect may vary considerably, even in pumps designed for the same performance.

A typical example of how increases in sealing-gap width affected the performance of a particular pump is presented in Fig. 4. The unit had a specific speed of about 2,000. When the radial clearance, b, was 0.012 in., the pump delivered 660 gal/min against a head of 47 ft. When b was 0.048 in., the flowrate at the same head dropped to 580 gal/min. Finally, when b reached 0.144 in., the flowrate fell to about 75 gal/min.

For relatively small increases in the width of the sealing gap, the effects of leakage on NPSH (net positive suction head) requirements are similar to the effects on the head-capacity curve. The same NPSH will be required for a flowrate equal to the original flow minus the increase in leakage flow, as illustrated in Fig. 5.

In the case of larger clearances, however, the liquid leaking through the

TABLE — COMMON MATERIALS LISTED IN ORDER OF INCREASING RESISTANCE TO WEAR

Aluminum and some of its alloys
Plastics (with the exception of some specialty materials)
Cast iron
Cast steel
Certain varieties of bronze
Manganese steels
Low-alloy chrome steels
High-alloy chrome steels
Certain varieties of rubber (used in conjunction with stainless steel for the mating parts)
Certain ceramics
Certain specialty alloys
Certain carbides

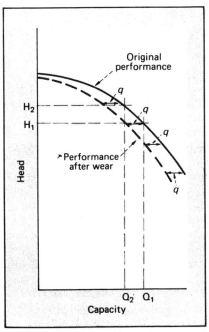

FIGURE 2. The shape of the gap between wear surfaces affects wear resistance

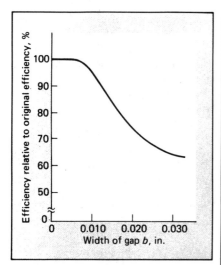

FIGURE 3. Wear causes increased leakage and changes the head-capacity curve

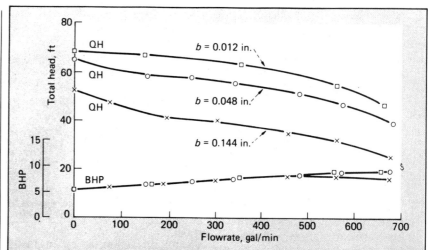

FIGURE 4. As the width of the gap increases, pump efficiency drops

sealing gap is able to completely change the flow pattern of the liquid entering the impeller. This usually affects the NPSH requirements in an unpredictable manner. In the author's experience, such increases in gap width produced higher NPSH requirements than would have been expected based on Fig. 5.

Changes in surface roughness

Depending on the material of construction and the properties of the liquid being pumped, the roughness of the flowpaths can also change over time. In some instances, the channels may acquire a smooth polish, and in others they may become roughened. Both of these changes can significantly affect pump performance.

Changes in the roughness of the casing affect performance differently than do changes in the roughness of the impeller. An increase in casing roughness usually reduces both the total head and the efficiency. However, changes in the surface roughness of the impeller produce more-complex results. First, one must distinguish between changes in the roughness of the inner flowpaths of the impeller and changes in the roughness of the outer shroud surfaces. Rougher inner flowpaths generally produce a drop in both efficiency and head. Rougher shrouds, on the other hand, often produce significantly *higher* heads, and generally result in lower efficiencies; however, in a few cases where the pump's specific speed has been very low (i.e., N_s =

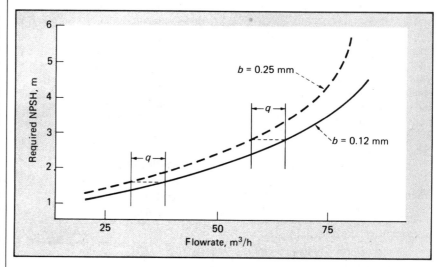

FIGURE 5. Performance of a pump operating at 1,760 rpm, as gap inceased

420–480), rougher shrouds have resulted in increased efficiencies and heads.

Although no formal research has been done on the effects of roughness on a pump's NPSH requirements (to the author's knowledge), visual observations of flow in pumps having transparent casings have revealed that any roughness or sharp projection usually causes intense local cavitation. To suppress this, it is often necessary to increase the available NPSH many times above that required to suppress cavitation in the absence of rough spots.

Changes in flowpath size

The dimensions of the pump's flowpaths may change over time due to abrasion or erosion, which usually increase the pathways' dimensions, or to scale, rust or sedimentation, which usually reduce the size of the pathways. The latter is particularly apt to occur in pumps operating intermittently.

As a general rule, an increase in the size of the flowpaths will increase the flowrate at a given head. Conversely, a reduction in the size of the flowpaths will reduce the flowrate.

The author

S. Yedidiah, a consulting engineer specializing in centrifugal pumps, can be reached at 89 Oakridge Rd., West Orange, NJ 07052; tel. 201-731-6293. He has been active in this field since 1938, and has worked for a number of pump manufacturers. He has published many articles, and is the author of "Centrifugal Pump Problems — Causes and Cures," published by PennWell Books. He is a member of the American Soc. of Mechanical Engineers, and is a registered professional engineer.

DESIGNING CENTRIFUGAL PUMP SYSTEMS

Steven M. Fischer, The Badger Co.

To properly specify the process requirements for a pump, the design engineer must explore the full range of operations the pump will be expected to perform. Considering only the normal operating case could lead to disappointing performance at either higher- or lower-than-normal flowrates.

The full range of anticipated operations are analyzed in this method for designing centrifugal pump systems to properly determine pump and control-valve hydraulic requirements. The method can also help minimize overdesign by revealing the effects imposed by large contingencies in flowrates, even when control-valve differentials are minimal.

The process calculations can be incorporated into a computer program, to enhance the analysis and consistency of designs. By substituting actual pump-curve data and piping-and-equipment pressure-drops for assumed values, the process design can be developed into a detailed design, to confirm the operability of a system.

The operating cases

For most processes, at least three operating cases should be examined, in order to ensure the operability of a pump system. These cases are: normal, rated, and alternates.

The normal flow case usually refers to the nameplate, steady state design capacity of the plant. It is often the point at which the process is expected to operate most of the time, and, therefore, the point at which the process designer should try to optimize the efficiency of the system.

Rated flows are normally established to provide some measure of contingency against variations in capacity resulting from changes or uncertainties in feedstocks, catalyst yields, separations, tray efficiencies, level swings, process upsets, etc. If a process is only expected to operate over a narrow range of flowrates, little or no contingency is required in flow capacity, and rated flow can equal normal flow.

When a rated flow is established for one of the foregoing reasons, it follows that the process is intended to operate in that condition. Therefore, the process designer must be certain that the pump can develop the necessary head to achieve the required rated flow. Only by doing the necessary hydraulic calculations will the designer know if there will be sufficient pressure drop available in the control valve (or valves) to compensate for the additional frictional pressure drop created at the rated flowrate.

Alternate flow cases need to be investigated to assure the flexibility of the design for any other operating cases, such as turndown operations.

Turndown cases are important because of the additional pressure drop that the control valves will have to absorb due to the combination of higher pump head and lower frictional pressure drop at reduced flowrates. Critical in such a case is that the control valves be able to maintain control of the process without chattering due to the excessive pressure drop available. Fig. 1 illustrates typical head vs. flow relationships for the various operating cases.

Design approach

First, the rated head and flow requirements are determined in order to select the pump. Next, the remaining flow cases are evaluated on a hydraulically consistent basis, using the same pump curve.

For revamp designs, curves may be available. For new designs, assumed or generalized pump curves, such as in Fig. 2, may have to be used. The Fig. 2

Originally published February 16, 1987

curves are characteristic of centrifugal pumps whose heads rise continuously to shutoff.

In many companies, a maximum allowable rise in head of a centrifugal pump at shutoff is prescribed so as to limit the design pressures of piping and equipment that may be blocked in. In such instances, the design engineer need only pick the curve that is consistent with his company's normal design practices, typically 20% or 25%, for use in the subsequent hydraulic calculations.

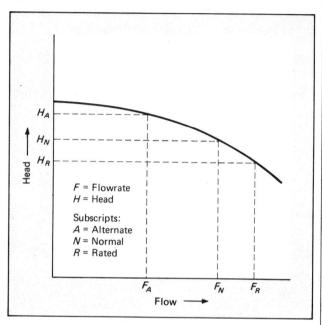

Figure 1 — Head vs. flow relationships for three operating cases

Usually, the allowable pressure drops in equipment and piping are known or specified for the normal flow case. To develop the hydraulics for the rated and alternate flow cases that are consistent with that for the normal flow case, the designer needs some way of estimating the frictional pressure drops that correspond to the flowrates for the other cases.

The frictional pressure drop at a second flow condition can be calculated if the drop is known at one flow condition, via a correlation based on the Darcy equation:

$$\Delta P_2 = \Delta P_1 (V_2/V_1)^2$$

Here, ΔP = frictional pressure drop, and V = velocity or flowrate.

The following assumptions pertain to this equation:
1. Only the flowrates are different. Temperature and composition remain the same. This eliminates the need to correct for changes in fluid properties.
2. Flow is turbulent throughout.
3. The friction factor, f, remains constant over the range of expected flowrates. Although f does vary with flow, the change is small, especially with well developed turbulent flow, relative to the effect that flowrate has on velocity, which is squared.
4. The frictional drop through equipment can be treated in the same manner as that through piping.

Sizing pumps for branched-flow systems

The flow required of a pump will be the sum of the simultaneous flows through the branches of a piping system. Branches may require different rated vs. normal flow capacities because of specific process requirements. For example, a distillation-column-overhead pump that handles both reflux and net product flow might need to be sized for an additional 20% reflux but only 10% additional net flow. The actual rated

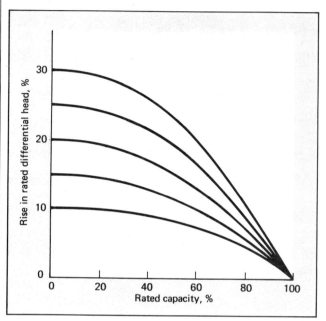

Figure 2 — Generalized curve may be substituted for specific one

flow would then be between 10% and 20% higher than the normal flow.

A piping section from which piping branches off is referred to as a common. Its frictional pressure drop is treated the same as that of any other pipe section. However, it contributes to the pressure drop of each of the branches.

After the flowrates have been determined for the commons and each branch — and, thus, for the pump — the frictional ΔPs for each case can be adjusted via the Darcy-based correlation, when flows differ from the normal case.

For pump selection, the design approach is to determine the rated head and flow requirements consistent with the operating conditions of the normal case. Therefore, the procedure starts with the rated calculations.

Example — Column bottoms pump

For the system shown in Fig. 3, determine the following: (a) pump head at rated flow; (b) control-valve pressure drops in each branch, for each case; (c) net positive suction head available (NPSHA) at the rated flow; and (d) maximum pump discharge pressure.

The following is given: a sketch of the pump system, and the flowrates *occurring simultaneously* in the branches, for each case (Fig. 3); estimated and allowable frictional pressure drops for commons, branches and equipment at normal flow (Table I); static heights and specific gravities

DESIGNING CENTRIFUGAL PUMP SYSTEMS

(Fig. 3); source and destination pressures (Table II); maximum allowable percent rise in head at shutoff; and fluid vapor pressure at the pumping temperature.

The calculation procedure:

1. For each operating case, determine the flowrates in each branch, in the commons, and at the pump, in order to be able to correct the frictional ΔP terms for the different flowrates. Because the volumetric flows in the branches could vary with temperature (one stream flow may be measured downstream of a cooler, for example), branch flows should be added on a weight basis to retain the material balance.

After all the flows have been determined, correct the frictional ΔP terms for the rated and alternate cases. For example, calculate the ΔP of the exchanger in Branch 2 at rated flow (150 gpm) if the allowable ΔP is 15.0 psi at normal flow (100 gpm). Thus, $\Delta P = 15.0 \ (150/100)^2 = 33.75$ psi.

To calculate the ΔP in the suction line at rated flow, determine the weight flowrates in this line for each case: Flowrate, lb/h = (500 lb/h/gpm) $\times \Sigma[(\text{gpm})(\text{sp.gr.})]_n$. Here, n refers to the branch numbers that contribute to the total flow.

Rated-case suction flow = 500 [(625)(0.75) + (150)(0.83) + (165)(0.80)] = 362,625 lb/h = 967 gpm.

Normal-case suction flow = 500 [(500)(0.75) + (100)(0.83) + (150)(0.80)] = 289,000 lb/h = 771 gpm.

Rated-case suction line ΔP = 2.0 (362,625/289,000)^2 = 3.1 psi.

All frictional ΔP terms are corrected in the same manner. Note, however, that the correction factors for all the branches, commons and pump sections are not necessarily the same, because ratios of rated-to-normal flowrates may be different, even for the same case.

2. Calculate the rated-case pump suction pressure, using the low liquid level (LLL) as the suction static height — i.e., pump suction pressure = source pressure + static head − friction losses = 64.7 + (18.5 × 0.75)/2.31 − 3.1 = 67.6 psia.

3. Calculate the net positive suction head: NPSHA = rated suction pressure − vapor pressure = 67.6 − 64.7 = 2.9 psi (or 8.9 ft).

4. Start the discharge calculations. The designer must allow a minimal control-valve differential for each branch, so that the process can be controlled, *even at rated flow*. Typically, this differential is based on a percentage of the frictional pressure drop (for instance, 10–20%, excluding the

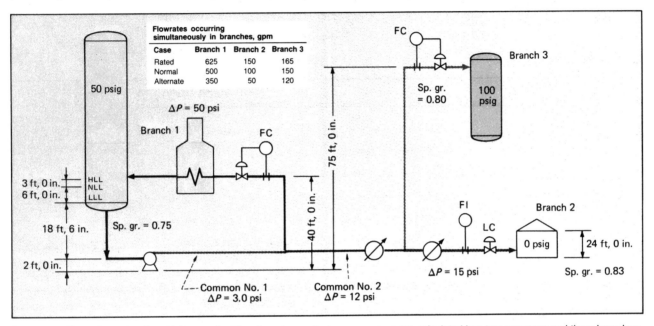

Figure 3 — Example system for which pump head and maximum discharge pressure are calculated has two commons and three branches

Table I — Normal-case frictional pressure drops, psi

	Branch 1	Branch 2	Branch 3
Line loss	5.0	15.0	8.0
ΔP, exchangers	0	15.0	0
ΔP, orifices	1.0	1.0	1.0
ΔP, dynamic head (min.)	2.0	2.0	2.0
ΔP, heater	50.0	0	0
ΔP, Common No. 1	3.0	3.0	3.0
ΔP, Common No. 2	0	12.0	12.0

control valve), or on an absolute minimum ΔP (5–10 psi, for example). For these guidelines, the destination pressures are assumed to remain constant in all the cases.

First, the pump discharge pressure is calculated individually for each branch, as follows: discharge pressure = destination pressure + static head + friction losses + control valve ΔP.

The branch having the highest discharge pressure will govern the pump discharge pressure, with the control valve pressure drops in the other branches being adjusted to fit the pump discharge pressure.

For Branch 1, calculate the total friction drop at rated

Table II — Given and calculated pressures and pressure drops in the three branches for the rated, normal and alternate flow cases

Pressures	Flow case								
	Rated			Normal			Alternate		
	Branch No.			Branch No.			Branch No.		
	1	2	3	1	2	3	1	2	3
Initial, psia	64.7			64.7			64.7		
+Static head, psi	6.0			8.0			8.0		
−Line loss, psi	3.1			2.0			1.0		
Pump suction, psia	67.6			70.7			71.7		
−Vapor pressure, psia	64.7								
NPSHA, psia (ft)	2.9 (8.9)								
Delivery, psia	64.7	14.7	114.7	64.7	14.7	114.7	64.7	14.7	114.7
Static head, psi	12.3	7.9	25.3	12.3	7.9	25.3	12.3	7.9	25.5
Line loss, psi	7.8	33.8	9.7	5.0	15.0	8.0	2.4	3.8	5.1
Control valve ΔP, psi	19.1	72.0	13.5	66.7	134.1	38.7	107.9	176.0	60.7
Exchanger ΔP, psi	0	33.8	0	0	15.0	0	0	3.8	0
Orifice ΔP, psi	1.6	2.3	1.2	1.0	1.0	1.0	0.5	0.2	0.6
Dynamic head ΔP, psi (2 psi minimum)	3.1	3.1	3.1	2.0	2.0	2.0	1.0	1.0	1.0
Heater ΔP, psi	78.1	0	0	50.0	0	0	24.5	0	0
Common No. 1 ΔP, psi	4.7	4.7	4.7	3.0	3.0	3.0	1.4	1.4	1.4
Common No. 2 ΔP, psi	0	19.2	19.2	0	12.0	12.0	0	5.5	5.5
Pump discharge, psia	191.4	191.4	191.4	204.7	204.7	204.7	214.3	214.3	214.3
−Suction, psia	67.6			70.7			71.7		
Pump differential, psi	123.8			134.0			142.6		
Pump differential, ft	381			413			439		

flow: $\Delta P = 7.8 + 0 + 1.6 + 3.1 + 78.1 + 4.7 + 0 = 95.3$ psi.

Next calculate the control-valve ΔP; the 20%, rather than the 10-psi minimum, rule governs here. Thus, the Branch 1 control valve minimum ΔP is 19.1 psi (95.3 psi × 20%).

The minimum pressure drops for the control valves in the other branches are calculated similarly.

In this example, Branch 1 governs the pump discharge pressure. The frictional pressure drop in Branch 2 = 33.8 + 33.8 + 2.2 + 3.1 + 0 + 4.7 + 19.2 = 96.8 psi. Therefore, the control valve ΔP would have been = 96.8 × 0.20 = 19.4 psi, and the pump discharge pressure would have been = 96.8 + 19.4 + 14.7 + 7.9 = 138.8 psia. However, because this is less than 191.4 psia discharge pressure calculated for Branch 1, the recalculated control valve ΔP for Branch 2 = 191.4 − 14.7 − 7.9 − 96.8 = 72.0 psi.

The pressure drops of the remaining branches are calculated similarly. Note that the branch having the control valve with the highest ΔP is not necessarily the governing branch.

5. Calculate the pump differential at rated flow. Pump differential = 191.4 − 67.6 = 123.8 psi (or 381 ft).

6. Calculate the maximum pump discharge pressure. Add the maximum pump differential pressure to the maximum corresponding pump suction pressure. The maximum pump suction pressure will occur when the vessel is being relieved, the liquid is at the high level (HLL), and the pump is shut off (i.e., maximum head and no frictional ΔP in the line).

For this example, the maximum rise in head at shutoff is taken to be 20% of the rated head, and the source pressure at relief is taken to be 65 psig (79.7 psia). Therefore, the maximum suction pressure = 79.7 + [(18.5 + 6.0 + 3.0)(0.75)]/2.31 − 0 = 88.6 psia. The maximum pump differential = 1.20 × 123.8 = 148.6 psi, and the maximum discharge pressure = 88.6 + 148.6 = 237.2 psia.

7. Calculate the normal-case hydraulics. For this case, no corrections to the friction terms are required. Calculate the normal suction pressure, using the static height at the normal liquid level (NLL).

8. Calculate the normal-case pump discharge pressure. Adjust the pump differential pressure for the lower pump flow, and add this to the normal case pump suction pressure. The normal pump flow = (771/967)(100) = 79.7% of the rated pump flow. Using the 20% maximum head rise curve in Fig. 2, find the new pump differential at normal flow to be 8.2% higher than the rated differential: Normal differential = 1.082 × 123.8 = 134.0 psi. Therefore, the normal pump discharge pressure = 70.7 + 134.0 = 204.7 psia.

9. Calculate the control valve pressure drops of each branch for the normal flow case. Subtract the respective destination pressure, static head and frictional pressure drop from the normal pump discharge pressure.

10. Calculate the hydraulics for the alternate case in the same manner as for the normal case, except for again correcting the frictional pressure drops for differences in flowrates, as was done for the rated-case flow.

The calculations are summarized in Table II.

The author

Steven M. Fischer was a senior process design engineer with Badger Engineers, Inc., when he wrote this article. He is now a senior process engineer with Champlin Petroleum Co., responsible for the heavy-ends processing units of the company's refinery at Corpus Christi, Texas (P.O. Box 9176, TX 78469; tel.: 512-887-3344). Prior to joining Badger Engineers, he had been a senior process design engineer with UOP Inc.'s Process Division. He holds a B.S.Ch.E. degree from Tulane University.

PUMP BYPASSES
NOW MORE IMPORTANT

Irving Taylor, Consultant, Bechtel, Inc.

Discoveries during the past two decades show that it is important to install automatic bypasses on many large centrifugal pumps. Else, serious problems may arise.

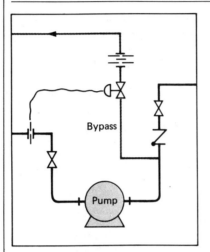

For all industrial centrifugal pumps (except boiler-feed pumps), automatic bypasses were usually omitted until 15 or 20 years ago. For industrial positive-displacement pumps, however, full-size bypasses have always been essential in the form of safety relief valves. These protect the pumps from overpressure, and also protect the downstream equipment. (The small-size bypasses previously required for boiler-feed pumps will be discussed in a list of "reasons why" below.)

Bypasses — when and why

Reciprocating and rotary positive-displacement pumps can be considered as a group, and have just one reason for always needing a relief-valve type of bypass (see Table I). Centrifugal pumps, regenerative-turbine-type pumps, and power-recovery turbines are also briefly covered in that tabulation.

For centrifugal pumps, there are now two primary reasons and about eight more occasional ones why some of them might need a bypass (Fig. 1). But before listing all ten reasons, there is one basic question to think about — one that most affects the need for a bypass. This is:

What is the expected frequency, and extent, of periods of low-flow operation — not forgetting possible operation at shutoff (zero delivery to the discharge system) when a downstream controlling valve is closed? (Very short "one minute" periods of low flow immediately after pump startups can usually be tolerated, except for axial-flow pumps and a few other types.)

Unfortunately, pump data sheets seldom show enough notes or remarks about possible lower-than-normal flowrates. Diligent pump application-engineers eventually develop suspicions about some pumps' service conditions. They may have to confer with process and project engineers (and even the plant's operators) to discover the few rare, but expected, situations in which continuous low flows could occur. For instance, a level-control valve in a pump discharge line can be expected to go fully closed occasionally, for more than just a minute.

Ten reasons for bypasses

Here are ten reasons why an industrial centrifugal pump may sometimes need a bypass:

1. To prevent excessive temperature-rise in the pump at very low flowrates. This had been especially necessary for boiler-feed pumps. Now, an automatic modulating minimum-flow bypass is necessary for almost any moderate or high-pressure centrifugal pump that will have its delivery automatically controlled (to maintain a liquid level or a process temperature), rather than being flow-controlled. The bypass will prevent excessive pump temperature-rises. Usually, less than 10% of normal flow is enough for a "pump thermal-control" bypass. (For very small pumps, below 25 gpm or so, a continuous bypass is

satisfactory; when the pump runs, the bypass flows regardless of what the delivered discharge-line flowrate is.)

2. To avoid unstable flow conditions in certain pumps — that is, to avoid pumping below the capacity onset-point of impeller recirculations, at which point vortex-type cavitation, low-frequency pulsations, and vibrations are all likely possibilities.

The pumps that are the most susceptible to unstable flow are those (a) with a high, suction specific-speed index-number (N_{ss}) — say, over 11,000 from Eq. (1), (using U.S. gpm

3. To prevent pumping at a capacity lower than the peak-point of a "nonrising to shutoff" head–capacity pump curve. In high-energy pumps, this can cause surging pulsations and pipe-shakings that are similar to centrifugal-compressor surgings.

4. To reduce shaft and bearing loads so as to prevent failures due to radial thrust in some older double-suction pumps having a single-volute design, especially if the pump head exceeds 200 ft. Again, low-flow operation, whether prolonged or repeated too often, is the culprit.

Table I — Pump bypasses for reciprocating, rotary and centrifugal pumps: when, why, where and how much?

Pump type	When required?	Why?	Where?	How much flow?
Reciprocating Direct acting Power pumps	Seldom Almost always	To prevent overpressuring the pump or the piping. (Usually use relief valves.)	Bypasses should branch from each pump's discharge line, upstream of the first valve, whether it be a check valve or a block valve	Full flow, 100%
Rotary	Almost always, unless built in			Full flow, 100%
Peripheral, regenerative turbine (semi-positive displacement)	Frequently			Judgment required for each application
Centrifugal	Only sometimes	Ten possible reasons (see list in article)		Judgment required (see text under "Minimum continuous stable flow").
HPRTs (hydraulic power-recovery turbines)	Almost always	To maintain process flow automatically	Branched from turbine's inlet line	Full flow, plus; >100%

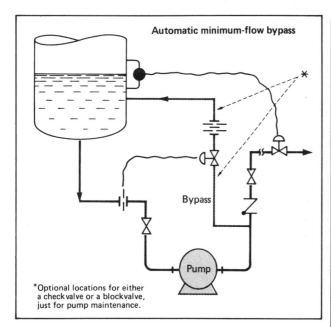

Figure 1 — Diagram of automatic, minimum-flow pump bypass

and ft units), and (b) high-energy pumps — say, over 200 m (or 650 ft) of head per stage, and over 225 kW per stage.

$$N_{ss} = \frac{\text{RPM} \times (\text{flowrate per impeller eye})^{1/2}}{(\text{NPSH required})^{3/4}} \quad (1)$$

Figure 2 — A typical bypass control valve (this has piston actuator)

5. To allow a few months of prolonged low-flow pumping during plant commissioning (especially for nuclear plants), or to permit repetition of slow warmup procedures in any plant.

6. To prevent damage to downstream equipment in the discharge system that cannot stand pressures higher than the pump's usual discharge pressure. (Increases of specific gravity or suction pressure, with a steep curve, might endanger equipment; the bypass valve might be a relief valve.)

7. During pump startups: (a) to prevent overloading the driver on axial-flow pumps; (b) to slowly fill an empty discharge-piping system. (A nonautomatic bypass valve should usually suffice.)

8. In a pump that is to deliver to two or more discharge branches at different times, with one of them at a much lower than normal flowrate. (To prevent pump trouble due to any of the previous reasons, see the "Pump Handbook," by Karassik, and others, 2nd ed., Section 8.2, Branch-Line Pumping Systems, Fig. 6).

9. To prevent air-binding (loss of prime) in some bulk-station and airport pumps that transfer hydrocarbon fuels (gasoline, kerosene, jet fuel, and some solvents), and that may be throttled to low flows or to shutoff while running. When these fuels happen to contain dissolved air (not always the case), the air can easily be cavitated out of solution by a centrifugal pump at low flows. Since the redissolving of air is very slow (not like the immediate collapses of true vapor cavities), recirculation at the impeller eye will accumulate enough air to block the throughflow and reduce the head developed. Then the discharge check valve can close and the running pump may not deliver at all until it is stopped and vented.

10. Forward-flow bypasses differ from the other situations. They are needed for multiple pipeline pumps when these are to operate in series.

If one pump should stop, the others must still be supplied with liquid to keep on pumping. Such bypasses have a line-size check valve to permit forward flow automatically when required.

Note: Warming lines for standby hot pumps are not really pump bypasses., they are only discharge-check valve bypasses, passing a tiny flow backward through the nonrunning pump.

Need for hydraulically stable pumping

Reason No. 2 of the above list has really come to the fore in the last 20 years. There was a spate of puzzling pump problems in the 1960s and early 1970s — mostly affecting high-head and high-horsepower pumps that were handling water. Some petroleum refinery pumps handling lean carbonate solutions also were afflicted. Excessive pulsations and vibrations occurred at low flowrates. Even unexplainable cavitation noises and some cavitation-erosion occurred at low flows. A few of these same pumps suffered breakages of impeller shrouds and pump shafts.

I believe that some Japanese and British pump researchers were the first to set up tests for visual flow studies of cavitation at pump impeller inlets, and to see and photograph the different shapes of cavitation "clouds", prerotations and backflows, through a transparent window at a pump inlet.

In 1960, Minami, and others, published a paper that included a landmark discovery about cavitation in centrifugal pumps. There were four sets of pump NSPHR (net positive suction head required) curves, for tests of four different impellers in one pump. Each set compared the measured NPSH required to suppress all cavitation (no incipient clouds or vapor bubbles observed), against the traditional curves of

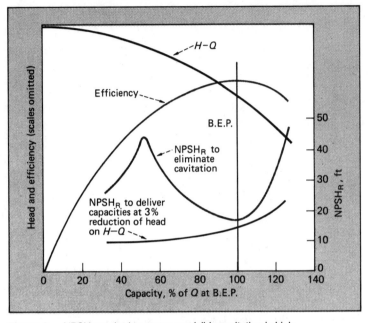

Figure 3 — NPSH required to suppress visible cavitation is high

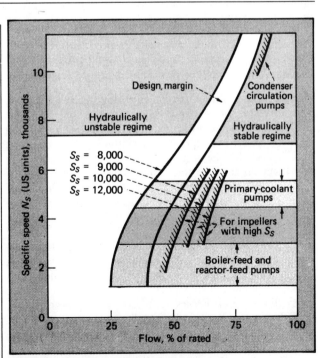

Figure 4 — A chart by Makay [10] showing useful operating ranges for stable pumping

NPSH required to prevent more than a 3% reduction in pump first-stage head. Fig. 3 shows a typical rendition of the comparison found in all four sets of the Minami curves. Compare the shape of the two curves, one for incipient cavitation visually determined under strobelight, and one determined for a 3% pump first-stage head-reduction. (Less than 3% makes for difficulties in getting repeatable test results.)

The previously overlooked amounts of cavitation occurring between incipience and performance impairment do not index-number. Some of us who had been against a wider use of this N_{ss} number were won over. Fraser showed that N_{ss} is one of the principal guides for deciding on a pump's minimum continuous flow. He did give us some modifying factors with his two figures showing recirculation-flow percentages. These should not be overlooked.

Bypasses — where?

Bypasses sometimes have been installed wrongly, so that some high-energy pumps suffered one of the worst types of

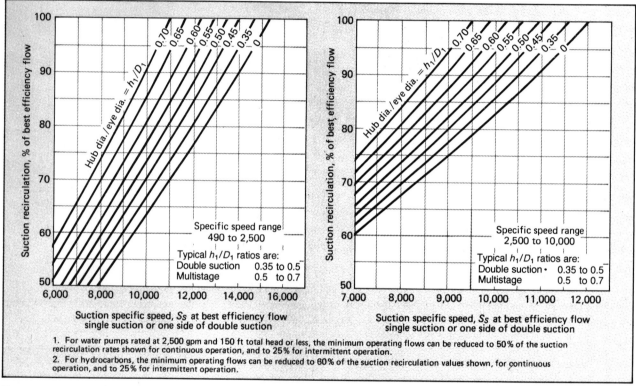

Figure 5 — Thse graphs for suction recirculations, by Warren Fraser [11] also give guidance on suggested minimum flows

always cause cavitation-erosion and other damage in pumps, but sometimes they can, and they have done so!

It came as a big hydraulics-surprise to learn that the required NPSH to suppress all cavitation at 40 to 60% of a pump's best-efficiency-point flowrate (b.e.p.) could be two to five times as much as was thought to be required just to meet head and capacity guarantees at rated flow. The curves of incipient cavitation NPSHR (that is, NPSH required to keep the cavitation clouds from starting to be seen) were very different.

Later, it was surprising to learn that those new "low flow" cavitations, along with vortices and partial backflows at the impeller inlet could make instabilities strong enough to shake the pump and the piping!

In 1981, Warren Fraser of Worthington presented and published his explanation for impeller recirculations — both at inlet and at impeller outlet — and his calculation methods to determine, for almost any centrifugal pump, at what percent of b.e.p. flowrate such recirculations would occur. He also clarified a new use for the suction specific-speed failure — seizure! Suction supply vessels have been exploded as a result of wrong bypass branch-off locations!

Automatic bypasses should branch off from *each* pump's discharge line *upstream of the first valve*, whether it be a check valve, a block valve, or control valve (except for those "forward flow" bypasses in Reason No. 10 above, and some nonautomatic ones.) Note the emphasis on the first italicized word, "each". There is a persistent erroneous assumption that a single, common, automatic, minimum-flow bypass can protect two or more pumps in the same service, such as a pump and its standby spare pump. Although a common bypass may have worked well enough for pairs of constant-speed pumps, both motor-driven, there are still two dangers:

1. If one or more of the pumps that has a common bypass branched off downstream of the check valves has a variable-speed drive, such as a steam turbine, the speed setting may not always equal the other pump's running speed.

When it is desirable to run both pumps at once, the speeds may differ enough so that the higher-speed pump will develop enough pressure to hold closed the check valve of the lower-speed pump. In such a case, the lower-speed pump can then overheat rapidly and fail, since the common bypass cannot protect it.

2. Far worse for common bypasses is what can happen if a running pump is tripped or stops (say, due to a power failure), and the other pump for any reason cannot be started. Then the common bypass automatically opens and allows the discharge system under high pressure to spill back into the suction vessel. At least one such vessel exploded violently due to the "misplaced check valve," or the wrongly placed bypass branch-off location! Even the non-common individual bypasses, if branched off downstream of the discharge check valves could cause such a danger.

See Fig. 1 for a simple diagram of an automatic minimum-flow bypass. The bypass should return to the suction supply-vessel whenever possible. If this point is too far away, it should return through a cooler or to a point in the long suction line distant enough to prevent overheating the circulating liquid.

Table II — Earlier recommended minimum flows for high-energy pumps are still valid

Liquid and impeller type	Actual ratio, $NPSH_A/NPSH_{R3PC}$	Recommended minimum flow, % of B.E.P. flow
Hydrocarbon, single- or double-suction	1.1 and higher	25*
Water or water solution, single suction	2 or less	35*
	2.5	30*
	3 or more	25*
Water or water solution, double suction	2 or less	70*
	2.5	60*
	3	50*

*Or manufacturer's recommendation, whichever is higher.

How much bypass flow?

In 1976, the writer presented an ASME paper [*3*] showing a short table of seven recommended minimum flows for high-energy pumps (see Table II). Those minimums ranged from 25 to 70% of b.e.p flow.

Since then, several others have published graphical charts that can be used to suggest suitable minimum flows (see Figs. 4 and 5 for some samples). The approximate sequence of authors was : W. Eadie, E. Makay, W. Fraser (2 curves), R M. Dubner and C. Heald (2 curves).

Bypasses capable of passing a bit more than full-flow are also essential for hydraulic power-recovery turbines (HPRTs), which are usually either reverse centrifugal pumps or Francis-impeller turbines (see Table I).

Minimum continuous stable flow

The main consideration about need for a bypass is the expected frequency and extent of periods of low-flow operation, but there are other considerations affecting the selection of *how much* bypass. These include:

1. The specific speed region. Above 6,000 N_s (U.S. units), axial- and mixed-flow pumps will require much more flow for stability than radial-flow pumps below 6,000 N_s (see Fig. 2).

2. The suction specific-speed index number. Pumps with higher N_{ss} (or an S_s on Figs. 4 and 5) will require more flow to remain stable. (Hallam of Amoco showed that above 11,000 N_{ss}, failure frequency increased significantly.)

3. The size of the pump, the capacity required

4. The NPSH(available)/NPSHR ratio.

5. The impeller energy level — feet of head per stage.

6. The likelihood of impeller erosion by the pumpage, either by solids or by cavitation. For the latter, hydrocarbons cause less erosion than water or water solutions.

7. The possible use of insert devices in the pump that can preserve stable operation at low flowrates.

Conclusion

Do not assume from all the foregoing reasons that most centrifugal pumps are going to need bypasses. It will still be possible for the average industrial centrifugal pump to be operated, when necessary, at any point from 10 to 110% of the b.e.p. flowrate (one exception: some radial-bladed-impeller, "partial emission" centrifugal pumps).

The new importance of having a bypass is mostly directed to the relatively few centrifugal pumps that: (a) are larger, (b) for high heads, (c) require considerable horsepower, and (d) are found in applications where some continuous low-flow operations are to be expected.

References

1. Minami, others, Experimental Study on Cavitation in Centrifugal Pumps, *Bull. JSME*, Vol. 3, No. 9, pp. 19–35 (1960).
2. Grist, E., Nett Positive Suction Head Requirements for Avoidance of Unacceptable Cavitation-Erosion in Centrifugal Pumps, C163/74 Conference on Cavitation, Fluid Machinery Group, Heriot-Watt University, Edinburgh, Scotland, Institution of Mechanical Engineers (1974).
3. Taylor, I., Controls for High-Energy Centrifugal Pumps to Prevent Pulsation and Cavitation-Erosion, ASME booklet I00098, Centrifugal Compressor and Pump Stability, Stall and Surge, p. 29–35, Mar. 1976.
4. Massey, I. C., The Suction Instability Problem in Rotodynamic Pumps, Paper 4-1, International Conference on Pump and Turbine design and Development, National Engineering Laboratory East Kilbride, Scotland (1976).
5. Bush, others, Coping with Pump Progress: The Sources and Solutions of Centrifugal Pump Pulsations, Surges and Vibrations, Part II, *Worthington Pump World*, Vol. 2, No. 1 (1976).
6. Taylor, I., The Most Persistent Pump-Application Problem for Petroleum and Power Engineers, ASME paper No. 1977-Pet-5.
7. Okamura and Miyashiro, Cavitation in Centrifugal Pumps Operating at Low Capacities, ASME booklet H-000123, pp. 243 ff. (1978).
8. Eadie, W. S., Design and Application of Feedwater Pumping Equipment for Improved Availability in Cyclic Operation, ASME paper No. 1979-JPGC-Pwr-8.
9. Nelson, W. E., Pump Curves Can Be Deceptive, National Petroleum Refiners Assn. paper No. MC-80-7 (1980).
10. Makay, E., Feedpump Suction Is Performance Key, *Power*, p. 104, Sept. 1981.
11. Fraser, W. H., Recirculation in Centrifugal Pumps, presented at ASME winter annual meeting, Nov. 1981.
12. Fraser, W. H., Flow Recirculation in Centrifugal Pumps, Proc. 10th Turbomachinery Symp., sponsored by Texas A & M University, College Station, Tex., Dec. 1981.

The author

Irving Taylor, 1150 Keeler Ave., Berkeley, CA 94708, is now a consultant, following retirement from Bechtel Inc., San Francisco. Prior to joining Bechtel in 1960, he served as a turbomachinery section-head in the Engineering Dept. of The Lummus Co. for 18 years. His specialty is the application of pumps, compressors and turbines. He holds a degree in mechanical engineering from Cornell University, is a Life Member of the American Soc. of Mechanical Engineers and is a licensed professional engineer in three states.

Centrifugal pumps and system hydraulics

Centrifugal pumps and their associated liquid systems are pervasive in the chemical process industries. This report provides detailed information on pump performance, suction capabilities, viscosity effects, operation at off-design conditions, and energy conservation.

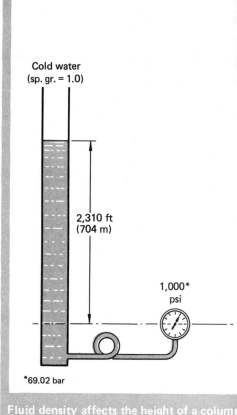

Fluid density affects the height of a column

Igor J. Karassik, Worthington Div., McGraw-Edison Co.

☐ Most processes in the chemical process industries (CPI) involve the transportation of liquids, or their transfer from one level of pressure or static energy to another.

The pump is the mechanical means for achieving this transport or transfer, and thus becomes an essential part of all processes. In turn, the growth and development of such processes are linked to the improvement of pumping equipment and to a better understanding of how pumps work and how they should be applied.

The centrifugal pump accounts for not less than 80% of the total pump production in the world because it is more suitable for handling large capacities of liquids than the positive-displacement pump. For this reason, we will examine the centrifugal pump exclusively in this article, and specifically try for a better understanding of centrifugal-pump and system hydraulics.

Head and system-head curves

Pumping is the addition of kinetic and potential energy to a liquid for the purpose of moving it from one point to another. This energy will cause the liquid to do work, such as flowing through a pipeline or rising to a higher level.

A centrifugal pump transforms mechanical energy from a rotating impeller into the kinetic and potential energy required. Although the centrifugal force developed depends on both the peripheral speed of the impeller and the density of the liquid, the amount of energy imparted per pound of liquid is independent of the density of the liquid. Therefore, for a given pump operating at a certain speed, and handling a definite volume of liquid, the energy applied and transferred to the liquid (in ft-lb/lb of liquid) is the same for any liquid, regardless of the density. (The only qualification to this statement is that the viscosity of the liquid does affect this energy, as we shall see later.) The pump head or energy in ft-lb/lb must, therefore, be expressed in feet.

Within the pumping system itself, we must remember that (1) head can be measured in various units, such as ft of liquid, pressure in psi, in. of mercury, etc.; (2) pressures and head readings can be in gage or absolute units (the difference between gage and absolute units is affected by the existing atmospheric pressure and therefore by the altitude); and (3) the pressure at any point in a system handling liquids must never be permitted to fall below the vapor pressure of the liquid.

A column of cold water 2.31 ft high will produce a pressure of 1 psi at its base. Thus, for water at ordinary ambient temperatures, any pressure calculated in pounds per square inch can be converted into an equivalent head in feet of water by multiplying by 2.31. For liquids other than cold water, the column of liquid equivalent to 1 psi pressure can be calculated by dividing 2.31 by the specific gravity of the liquid. The effect

Originally published October 4, 1982

CENTRIFUGAL PUMPS AND SYSTEM HYDRAULICS

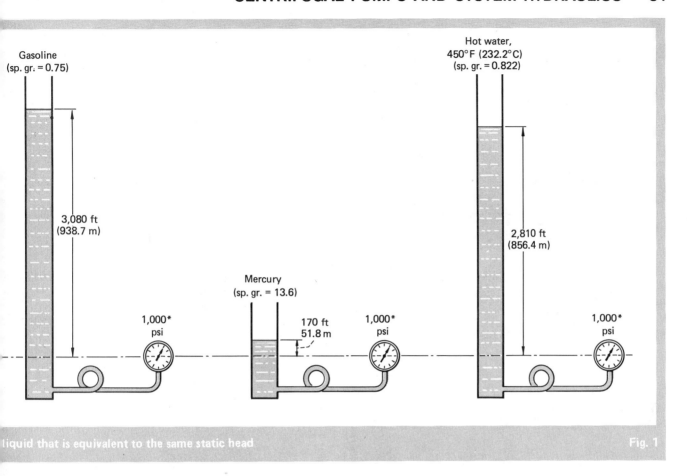

liquid that is equivalent to the same static head   Fig. 1

of specific gravity on the height of a column of various liquids for equal pressures is illustrated in Fig. 1. Formulas for the conversion of pressure and head data are given in Table I.

Fig. 2 illustrates the relationship between gage- and absolute-pressure readings. While it is usually feasible to work in terms of gage pressure, a complicated problem can occasionally be clarified by working entirely in terms of absolute pressure.

System head

In strict terms, a pump can only operate within a system. To deliver a given volume of liquid through that system, a pump must impart energy to the liquid, made up of the following components:
- Static head.
- Difference in pressures on liquid surfaces.
- Friction head.
- Entrance and exit losses.

Static head

The static head refers to a difference in elevation. Thus, the "total static head" of a system is the difference in elevation between the liquid levels at the discharge and the suction points of the pump (Fig. 3). The "static discharge head" is the difference in elevation between the discharge liquid level and the pump centerline. The "static suction head" is the difference in elevation between the suction liquid level and the pump centerline. If the static suction head has a negative value because the suction liquid level is below the pump centerline, it is usually spoken of as a "static suction lift." If either the suction or discharge liquid level is under a pressure other than atmospheric, this pressure can be considered either as part of the static head or separately as an addition to the static head.

Friction head

The friction head is the head (expressed in feet of the liquid being pumped) that is necessary to overcome the friction losses caused by flow of liquid through piping,

Equivalents for pressure and head — Table I

Gage pressure + Atmospheric pressure = Absolute pressure

U.S. Units	Metric Units
1 atmosphere = 14.7 psi	1 atm = 1.013 bar
1 atm = 34-ft column of cold water	1 atm = 1,013 mbar 1 atm = 10.33-m column of cold water
$\dfrac{34 \text{ ft}}{14.7 \text{ psi}} = 2.31 \text{ ft/psi}$	$\dfrac{10.33 \text{ m}}{1.013 \text{ bar}} = 10.2 \text{ m/bar}$
$\text{psi} = \dfrac{\text{head in ft}}{2.31} \times \text{sp. gr.}$	$\text{Pressure in bar} = \dfrac{\text{head in m}}{10.2} \times \text{sp. gr.}$
$\text{Head in ft} = \dfrac{\text{psi} \times 2.31}{\text{sp. gr.}}$	$\text{Head in m} = \dfrac{\text{bar} \times 10.2}{\text{sp. gr.}}$

52 CENTRIFUGAL PUMPS

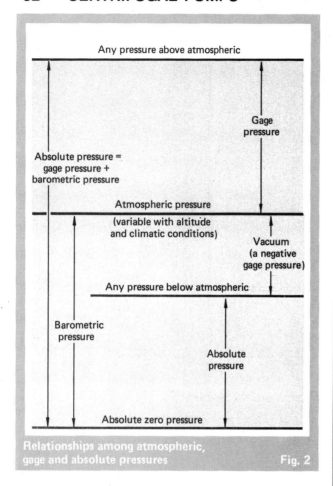

Relationships among atmospheric, gage and absolute pressures Fig. 2

valves, fittings and any other elements such as heat exchangers. These losses vary approximately as the square of the flow through the system. They also vary with the size, type and surface condition of the piping and fittings and with the character of the liquid pumped.

In calculating friction losses, we must consider that they will increase as the piping deteriorates with age. It is usual to base the losses on data established for average piping that is 10 or 15 years old. These data are readily available from sources such as the Hydraulic Institute Standards and the Pump Handbook [1,2].

Entrance and exit losses

If the supply of a pump originates in a reservoir, tank or intake chamber, losses occur at the point of connection of the suction piping to the source of supply. The magnitude of these losses depends on the design of the pipe entrance. A well-designed bell-mouth provides the lowest possible loss. Similarly, on the discharge side of the system where the discharge line terminates at some body of liquid, the velocity head of the liquid is entirely lost, and must be considered as part of the total friction losses of the system.

System-friction and system-head curves

As mentioned earlier, friction, entrance and exit losses vary approximately as the square of the flow through a system. For solving pumping problems, it is convenient to show the relationship between capacity and friction-head losses graphically. These losses are

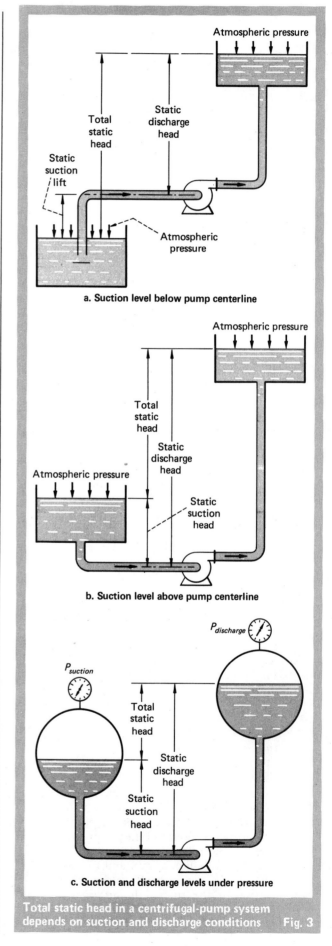

Total static head in a centrifugal-pump system depends on suction and discharge conditions Fig. 3

CENTRIFUGAL PUMPS AND SYSTEM HYDRAULICS

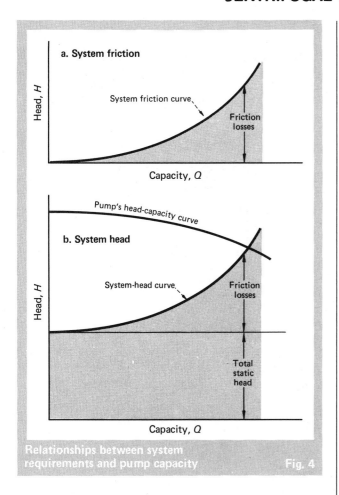

Relationships between system requirements and pump capacity — Fig. 4

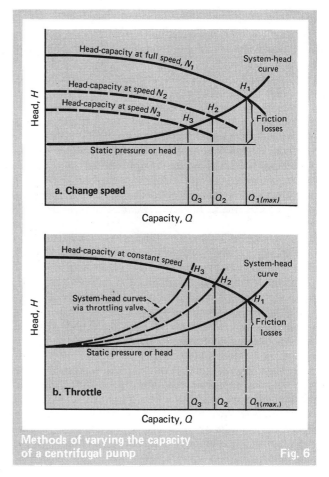

Methods of varying the capacity of a centrifugal pump — Fig. 6

therefore calculated at some predetermined flow, either expected or design, and then calculated for all other flows, using the square of the flow relationship. The resulting curve is called the system-friction curve, as shown in Fig. 4a.

When we combine the static heads, pressure differences and friction-head losses of any system and plot them against the capacity, the resulting curve (Fig. 4b) is called the system-head curve. Superimposing a pump head-capacity curve at constant speed on this system-head curve (Fig. 4b) will permit us to determine the capacity at the point where the two curves intersect. This is the capacity that will be delivered into the system by that pump at that particular speed.

For systems having varying static heads or pressure differences, it is possible to construct curves corresponding to the minimum and maximum conditions, as shown in Fig. 5. The corresponding intersections with the pump's head-capacity curve will then determine the minimum and maximum flows that the pump will deliver into the system.

Variations in desired flow

It is unusual for a system to require operation at a single fixed capacity. Generally, the process served by the centrifugal pump is variable in its demand. A given pump operating in a given system will only deliver that capacity corresponding to the intersection between the head-capacity and the system-head curves. In order to vary the capacity, it becomes necessary to change the shape of either one or both curves.

The pump head-capacity curve can be changed by operating the pump at variable speed (Fig. 6a). (For a guide to variable drives, see "Making the proper choice of adjustable-speed drives," p. 161.) Or the system-head curve can be altered by creating friction loss through a throttling valve (Fig. 6b).

Obviously, the difference between the total head de-

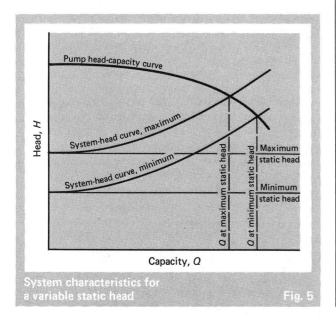

System characteristics for a variable static head — Fig. 5

54 CENTRIFUGAL PUMPS

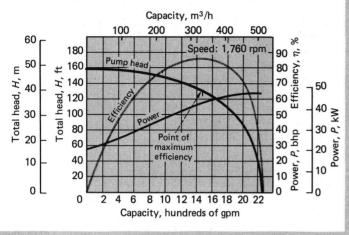

Performance characteristics for a double-suction centrifugal pump   Fig. 7

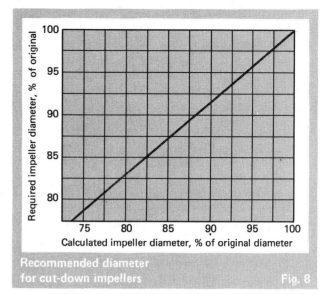

Recommended diameter for cut-down impellers   Fig. 8

veloped by the pump and the head required by the system-head curve represents wasted energy lost in the throttling process. On the other hand, the majority of centrifugal pumps today are driven by constant-speed squirrel-cage induction motors, and throttling the pump discharge is the only practical means of obtaining the desired variable capacity. As will be seen later in this article, the coming revolutionary move to variable-frequency motor drives is about to change these practices.

Affinity laws and rating curves

The hydraulic performance of a centrifugal pump involves three basic parameters: (1) the capacity (expressed in units of volume per unit of time such as gpm), (2) the total head (expressed in feet of the liquid pumped), and (3) the speed at which the pump runs (generally in rpm).

Normally, pump performance is presented in the form of curves such as shown in Fig. 7, with the head-capacity curve plotted at a fixed speed. The curve also shows the brake horsepower required at various flows, and the corresponding pump efficiency. The capacity at which the pump performs its function most efficiently is called the "b.e.p." or best efficiency point.

The useful work done by a pump is the weight of liquid pumped in a period of time, multiplied by the head developed by the pump, and is expressed in terms of horsepower, called water horsepower, WHP. It would be more correct to refer to WHP as liquid horsepower, which can be determined from:

$$WHP = \frac{QH(sp.\ gr.)}{3,960} \quad (1)$$

where: WHP = water horsepower; Q = pump capacity, gpm; and H = total head, ft.

The power required to drive the pump is the water horsepower divided by the pump efficiency, η. Hence, dividing Eq. (1) by η gives:

$$BHP = \frac{QH(sp.\ gr.)}{3,960\ \eta} \quad (2)$$

Affinity laws

The relationships that allow us to predict the performance of a pump for a speed other than that for known pump characteristics are referred to as the "affinity laws." When the speed is changed:

1. The capacity, Q, for any given point on the pump characteristics varies directly as the speed, n.
2. The head, H, varies as the square of the speed.
3. The brake horsepower, P, varies as the cube of the speed.

In other words, if Subscript 1 is assigned to the conditions under which the characteristics are known, while Subscript 2 denotes the conditions at some other speed, then:

$$\frac{Q_2}{Q_1} = \frac{n_2}{n_1}; \quad \frac{H_2}{H_1} = \left(\frac{n_2}{n_1}\right)^2; \quad \frac{P_2}{P_1} = \left(\frac{n_2}{n_1}\right)^3 \quad (3)$$

These relationships can be used safely for moderate changes in speed. Eq. (3) may not be as accurate for large speed changes.

Similar affinity laws exist for changes in impeller diameter, D, within reasonable limits of impeller cutdown. In other words:

$$\frac{Q_2}{Q_1} = \frac{D_2}{D_1}; \quad \frac{H_2}{H_1} = \left(\frac{D_2}{D_1}\right)^2; \quad \frac{P_2}{P_1} = \left(\frac{D_2}{D_1}\right)^3 \quad (4)$$

Some deviation from these laws occurs even with relatively modest cutdowns. Fig. 8 shows the recommended cutdown related to the theoretical cutdown.

Specific speed

The principle of dynamical similarity when applied to centrifugal pumps expresses the fact that two pumps geometrically similar to each other will have similar performance characteristics.

The term "specific speed" is the concept that links the three main parameters of the performance characteristics—capacity, head, and rotative speed—into a single term. The mathematical analysis used to establish the relationship between the specific speed of a pump and its operating characteristics need not concern us. In its basic form, the specific speed is an index number, expressed as:

$$N_s = \frac{n\sqrt{Q}}{H^{3/4}} \quad (5)$$

in which: N_s = specific speed; n = rotative speed, rpm; Q = capacity, gpm; and H = head, ft (head per stage for a multistage pump).

Eq. (5) remains unchanged whether the impeller is single-suction or double-suction. Therefore, it is customary when referring to a definite value of specific speed to mention the type of impeller.

While we could calculate the specific speed for any given operating condition of head and capacity, the definition of specific speed assumes that the head and capacity used in the equation refer to those at the best efficiency of the pump. The specific-speed number is independent of the rotative speed at which the pump is operating.

We cannot overemphasize the fact that "specific speed" is an index number—a concept quite similar to a "family name" that is useful in identifying various characteristics of a group. Just as the Browns, the Wilsons or the Smiths may be said to have certain hair or eye coloration, certain general common features, so do pumps of the same specific speed have a number of characteristics that distinguish them from pumps having other specific speeds.

For instance, the physical characteristics and the general outline of impeller profiles are intimately connected to their respective specific speeds. Thus, the value of the specific speed will immediately describe the approximate impeller shape, as shown in Fig. 9. Similarly, the specific speed of a given pump will be definitely reflected in the shape of the pump characteristic curves, as shown at the top of Fig. 9. While some variation in the shape of these curves can be made by changes in the design of the impeller and casing waterways, the variation that can be obtained without adversely affecting pump efficiency is relatively small.

Another parameter affected by the specific speed is the maximum efficiency obtainable from pump impellers of different specific speeds and different sizes, also indicated in Fig. 9.

Type characteristics for a pump

If the operating conditions for a pump at the design speed (that is the capacity, head, efficiency and power input at which the efficiency curve reaches its maximum) are taken as the 100% standard of comparison, the head-capacity, power-capacity and efficiency-capacity curves can all be plotted in terms of the percentage of their respective values at the capacity for maximum efficiency. Such a set of curves represents the "type characteristic" or "100%" curve of the pump.

The 100% curves of pumps having specific speeds of 2,000, 4,000 and 10,000 are shown in Fig. 10. These curves can be used to predict the approximate shape of a pump's characteristics once the specific speed of that pump is known. To avoid interpolation, the curves of Fig. 11 and 12 show the change in head and power in relation to specific speed for single-suction impellers. A double-suction impeller will have a type characteristic

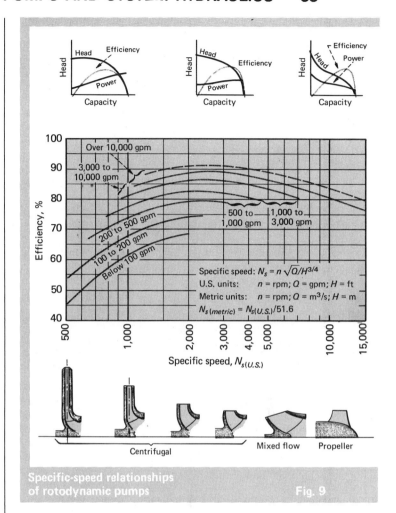

Specific-speed relationships of rotodynamic pumps — Fig. 9

approximating that of a single-suction impeller having a specific speed of $(1/2)^{1/2}$, or 70.7% that of the double-suction impeller.

Rating curves

Rating curves are commonly reproduced in pump bulletins and sales literature for standard lines of pumps. A rating curve for a centrifugal pump shows in a condensed form the possible range of applications of that pump at some rated speed for a range of impeller diameters (Fig. 13). A different chart is generally made available for each motor speed for a particular pump. In addition, the rating curves generally show a curve for the required net positive suction head (*NPSH*). To facilitate the selection of a pump, rating curves of an entire line of similar pumps are prepared.

Suction conditions

Most centrifugal-pump troubles occur on the suction side. Therefore, it is imperative to understand how to relate the suction capability of a centrifugal pump to the suction characteristics of the system in which it will operate.

When pumping liquids, the pressure at any point within the pump must never be permitted to fall below the vapor pressure of the liquid at the pumping temperature. There must always be sufficient energy available

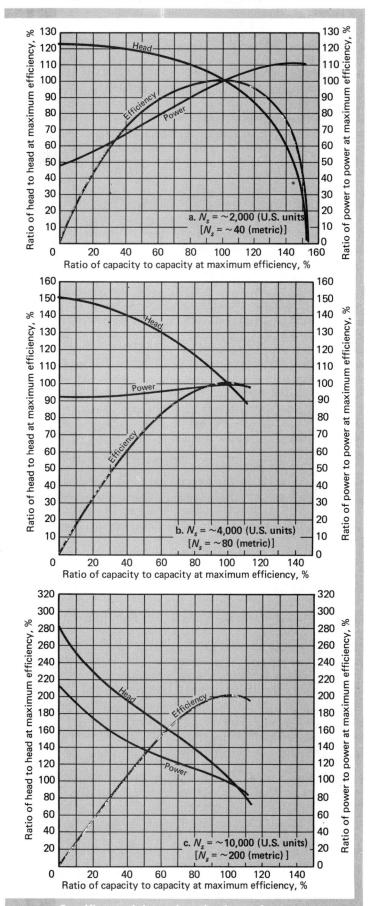

Specific speed determines the shape of a centrifugal pump's characteristics   Fig. 10

at the pump suction to get the liquid into the impeller and to overcome the losses between the pump-suction nozzle and the impeller inlet. At this point, the impeller vanes can act to add energy to the liquid.

Normally, pump performance is presented in the form of curves (Fig. 14), with the head-capacity curve for the pump plotted at a fixed speed. The curves also show the brake horsepower required at various flows, and the corresponding efficiency.

An additional characteristic of the pump is the (NPSH) required. This is the energy in feet of liquid head required at the pump suction over and above the vapor pressure of the liquid, to permit the pump to deliver a given capacity at a given speed.

Changes in available (NPSH) do not affect the pump performance as long as available (NPSH) remains above the value of required (NPSH). However, when available (NPSH) falls below the required value, the pump begins to cavitate and "works in the break," as shown in Fig. 14. The characteristics in solid lines are for values of (NPSH) available in excess of (NPSH) required. If available (NPSH) falls below that required—as for instance if at 1,800 gpm the (NPSH) available falls below 17 ft—cavitation starts, and the pump produces less head. Some increase in capacity occurs with a further reduction in head, until about 1,970 gpm is reached; then further reduction in head causes no increase in capacity (see curve with broken line).

Suction head and suction lift

As defined by the Standards of the Hydraulic Institute, the suction head, h_s, is the static head on the pump-suction line above the pump centerline, minus all friction-head losses for the capacity being considered (including entrance loss in the suction piping), plus any pressure (a vacuum being a negative pressure) existing in the suction supply.

Rather than express the suction head as a negative value, the term "suction lift" is normally used when the pump takes its suction from an open tank under atmospheric pressure. Because the suction lift is a negative suction head measured below atmospheric pressure, the total suction lift (symbol also h_s) is the sum of the static suction lift measured to the pump centerline and the friction-head losses as defined above. (It is sometimes advantageous to express both suction and discharge heads in absolute pressure, but usually it is more convenient to measure them above or below atmospheric pressure.)

A gage on the suction line to a pump, with its readings corrected to the pump centerline, measures the total suction head above atmospheric pressure minus the velocity head at the point of attachment. Because suction lift is a negative suction head, a vacuum gage will indicate the sum of the total suction lift and velocity head at the point of attachment.

The three most common suction-supply conditions are illustrated in Fig. 15.

Case I involves a suction supply under a pressure other than atmospheric, and located above the pump centerline. It includes all the components of the suction head, h_s. If h_s is to be expressed as a gage reading, and P_s is a partial vacuum, the vacuum expressed in feet of

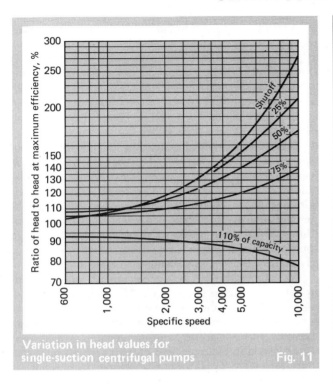

Variation in head values for single-suction centrifugal pumps — Fig. 11

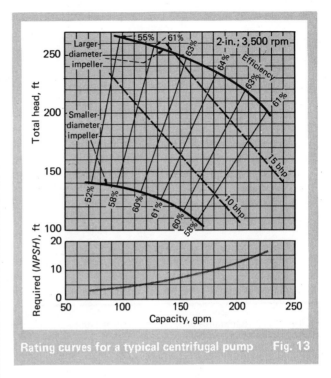

Rating curves for a typical centrifugal pump — Fig. 13

liquid would constitute a negative pressure head and carry a minus sign. If the pressure, P_s, is expressed in absolute pressure values, h_s will also be in absolute pressure values.

Case II involves a suction supply under atmospheric pressure, located above the pump centerline. Because the suction head (expressed as a gage value) has a P_s value of zero, the P_s value can be dropped from the formula given in Fig. 15.

Case III involves a suction supply under atmospheric pressure, located below the pump centerline. It is optional whether the suction head be expressed as a negative suction head or in positive values as a suction lift.

Because the source of supply is below the pump centerline (which is the datum line), S is a negative value. The suction-lift formula is the same as that for suction head, except that both sides have been multiplied by (-1). A gage attached to the pump suction flange, when corrected to the pump centerline, will register a partial vacuum or negative pressure.

To determine the suction head, it is therefore necessary to add the velocity head to this negative pressure algebraically; or, if it is desired to work in terms of a vacuum, the velocity head must be subtracted from the vacuum to obtain the suction lift.

For example, if the gage attached to the suction side

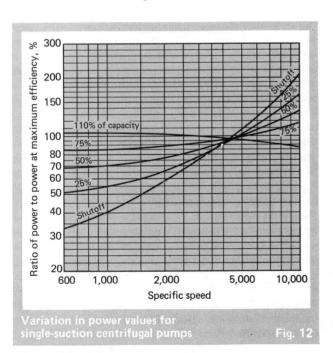

Variation in power values for single-suction centrifugal pumps — Fig. 12

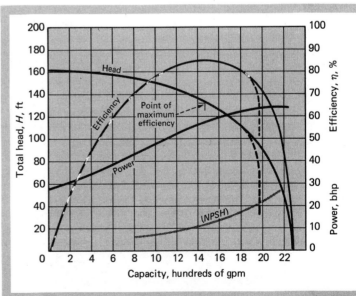

Performance characteristics for a centrifugal pump — Fig. 14

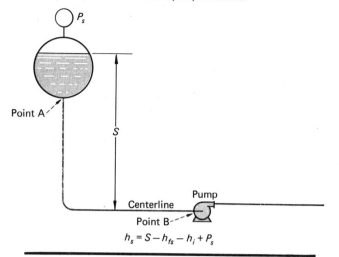

Case I—Suction from source under pressure other than atmospheric, and located above pump centerline.

$h_s = S - h_{fs} - h_i + P_s$

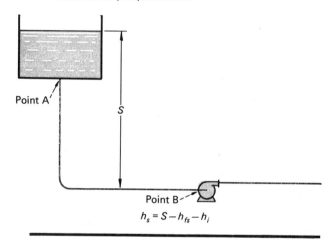

Case II—Suction from source under atmospheric pressure, and located above pump centerline.

$h_s = S - h_{fs} - h_i$

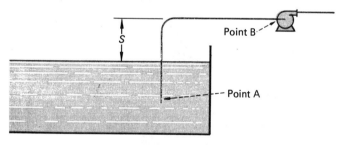

Case III—Suction from source under atmospheric pressure, and located below pump centerline.

$h_s \text{ (suction head)} = (-S) - h_{fs} - h_i$
$(-h_s)\text{(suction lift)} = S + h_{fs} + h_i$

h_i = Entrance loss at Point A
h_{fs} = Total friction loss from Point A to Point B
h_{vs} = Velocity head at Point B
h_{sg} (gage reading at Point B corrected to pump centerline) = $h_s - h_{vs}$
P_s, ft of liquid

How to determine suction head   Fig. 15

of a pump having a 6-in. line and pumping at a capacity of 1,000 gpm of cold water showed a vacuum of 6 in. Hg (equivalent to 6.8 ft of water), the velocity head at the gage attachment would be 2.0 ft of water, and the suction head would be $-6.8 + 2.0$, or -4.8 ft of water, or the suction lift would be $6.8 - 2.0$ or 4.8 ft of water.

Net positive suction head

The use of "permissible suction lift" or of "required suction head" has definite shortcomings. Either term can only be applied to water, because it refers to the energy of barometric pressure expressed in feet of water. Changes in barometric pressure, whether caused by differences in elevation above sea level or by climatic conditions, affect the value of these terms. Changes in pumping temperature also affect these values, since they affect the vapor pressure of the liquid.

For this reason, all references to suction conditions today are made in terms of (NPSH)—net positive suction head above the liquid vapor pressure.

Both suction head and vapor pressure should be expressed in feet of liquid being handled, and must both be expressed in either gage or absolute pressure units. A pump handling 62°F water (vapor pressure of 0.6 ft) at sea level, with a total suction lift of 0 ft, has an (NPSH) of $33.9 - 0.6$, or 33.3 ft, whereas one operating with a 15-ft total suction lift has an (NPSH) of $33.9 - 0.6 - 15$, or 18.3 ft.

A pump operating on suction lift will handle a certain maximum capacity of cold water without cavitation. The (NPSH) or amount of energy available at the suction nozzle of such a pump is the atmospheric pressure minus the sum of the suction lift and the vapor pressure of the water. To handle the same capacity with any other liquid, the same amount of energy must be available at the suction nozzle. Thus, for a liquid at its boiling point (in other words, under a pressure equivalent to the vapor pressure corresponding to its temperature), this energy has to exist entirely as a positive head. If the liquid is below its boiling point, the suction head required is reduced by the difference between the pressure existing in the liquid and the vapor pressure corresponding to the temperature.

It is necessary to differentiate between available net positive suction head, $(NPSH)_A$, and required net positive suction head, $(NPSH)_R$. The former, which is a characteristic of the system in which a centrifugal pump works, represents the difference between the existing absolute suction head and the vapor pressure at the prevailing temperature. The $(NPSH)_R$, which is a function of the pump design, represents the minimum required margin between the suction head and vapor pressure.

The manner in which $(NPSH)_A$ at a given capacity should be calculated for (1) a typical installation with a suction lift, (2) a pump taking its suction from a tank, and (3) a pump handling a liquid at the boiling point is demonstrated in Fig. 16.

Both $(NPSH)_A$ and $(NPSH)_R$ vary with capacity, as shown in Fig. 17. With a given static pressure or elevation difference at the suction side of a centrifugal pump, $(NPSH)_A$ is reduced at larger flows by the friction losses in the suction piping. On the other hand, $(NPSH)_R$, being a function of the velocities in the pump-suction

CENTRIFUGAL PUMPS AND SYSTEM HYDRAULICS

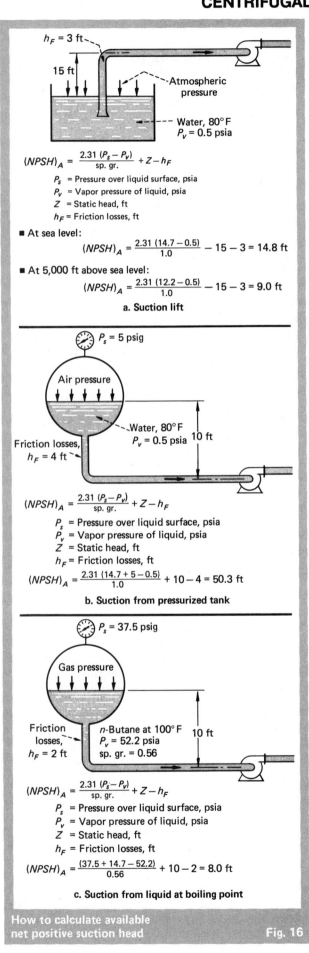

How to calculate available net positive suction head — Fig. 16

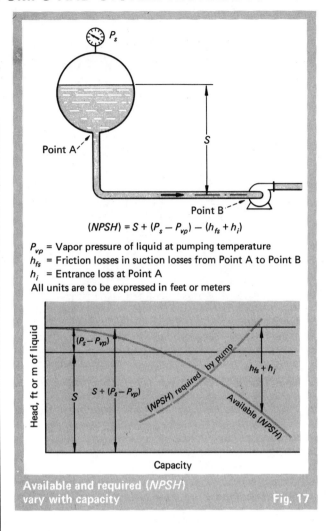

Available and required (NPSH) vary with capacity — Fig. 17

passages and at the inlet of the impeller, increases basically as the square of the capacity.

A great many factors—such as eye diameter, suction area of the impeller, shape and number of impeller vanes, area between these vanes, shaft and impeller hub diameter, impeller specific speed, and the shape of the suction passages—enter in some form or another into the determination of $(NPSH)_R$. Different designers may use different methods to produce an impeller that will perform satisfactorily with a specific value of $(NPSH)_R$. As a result, it is not recommended that users attempt to estimate $(NPSH)_R$ from the knowledge of just one or two of these factors. Instead, they should base their selections on data provided by pump manufacturers.

Specific-speed and suction limitations

Specific-speed-limit charts have been prepared and published by the Hydraulic Institute [1] for several types of pumps:
- Double-suction.
- Single-suction, with shaft through eye of impeller.
- Single-suction, overhung-impeller.
- Single-suction, mixed- and axial-flow.
- Hot-water, single-suction and double-suction.
- Condensate pumps, with shaft passing through eye of impeller.

One such chart, containing specific-speed limits for

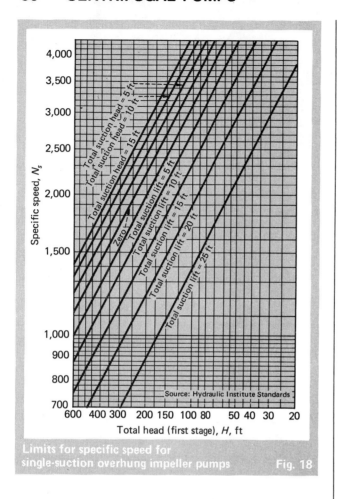

Limits for specific speed for single-suction overhung impeller pumps Fig. 18

single-suction overhung-impeller pumps, such as the ANSI pumps, is shown in Fig. 18.

It is important to remember that these charts are strictly empirical. In using them, it must be realized that pumps built for the limit allowed are not necessarily the best design for the intended service and that a lower-specific-speed type might be more economical.

It must also be realized that the design of individual pumps controls the application of the specific-speed limit for maximum head and suction conditions. For example, the maximum recommended specific speed for a double-suction, single-stage pump is 1,990 for a 200-ft total head and a 15-ft suction lift. It does not follow that all double-suction, single-stage pumps of 1,990 specific-speed type are suitable for operation at speeds that will cause them to develop a 200-ft total head (at maximum efficiency); nor that the pump, if suitable for operation at a 200-ft total head, is suitable for operation with a 15-ft suction lift; nor that a pump of this type operating against a 200-ft total head would on test be found capable of operating on only a 15-ft maximum suction lift.

These charts are intended to indicate only the maximum rotative speed for which experience has shown a centrifugal pump can be designed with assurance of reasonable and proper operation for the combination of operating conditions. The Hydraulic Institute Standards suction-limitation charts should be considered guidelines.

Nothing in the Hydraulic Institute Standards suction-limitation charts suggests that the specific speed indicated corresponds to the point of maximum efficiency. Yet this is the intended meaning. If a pump is applied for conditions near to its capacity at the best efficiency, there would be little error introduced by using the rated conditions to determine chart limitations. On the other hand, if the rated and best-efficiency conditions were to differ significantly, the chart recommendations would be found to apply only to the best efficiency point.

Suction specific speed

Application of the specific-speed-limit charts, as originally developed, had a very important shortcoming, i.e., satisfactory suction conditions were tied directly to the total head developed by the pump. The performance of an impeller from the point of view of cavitation cannot be affected too significantly by conditions existing at the impeller's discharge periphery. Yet, these conditions are the prime factors in determining the total head that the impeller will develop.

In other words, if an impeller exhibits certain suction characteristics, cutting down its diameter within reasonable limits, and thus reducing its head, should have no influence on its suction capabilities. Since the total head, H, is changed, a strict interpretation of the specific-speed-limit charts would indicate that unless the suction lift is to be proportionately altered, the maximum permissible specific speed must be changed.

This inconsistency was finally resolved by the development of the suction specific-speed concept. It is essentially an index number, descriptive of the suction characteristics of a given impeller, and is defined as:

$$S = \frac{n\sqrt{Q}}{(h_s)^{3/4}} \qquad (6)$$

where: S = suction specific speed; n = rotative speed, rpm; Q = flow, gpm. (For single-suction impellers, Q is the total flow. For double-suction impellers, Q is taken as one-half of the total flow.); and h_s = required (NPSH), ft.

The specific-speed-limit charts (such as in Fig. 18) have been revised several times since they were first adopted as a guideline for centrifugal-pump suction conditions. Unfortunately, they are still based on the erroneous concept that the total head developed by the pump plays a part in determining the maximum permissible rotative speed for a given set of suction conditions, despite the recognition of the concept of suction specific speed in the Hydraulic Institute Standards. The charts in the Standards today are based on values of S ranging from 7,480 to 10,690. These values vary within each chart, as well as from chart to chart. It is hoped that the charts will be revised and simplified so as to make them easier to use. They will then resemble the chart for the hot-water pump (see Fig. 19), in which the required (NPSH) can be read directly for any given flow at various speeds.

Cavitation and pump performance

Cavitation occurs when the absolute pressure within an impeller falls below the vapor pressure of the liquid, and bubbles of vapor are formed. These bubbles collapse further out along the impeller blades when they reach a region of higher pressure. The minimum required (NPSH) for a given capacity and a given pump speed is defined as that difference between the absolute suction head and the vapor pressure of the liquid pumped at the pumping temperature that is necessary to prevent cavitation.

Pump cavitation becomes evident when there is one or more of the following signs: noise, vibration, drop in the head-capacity and efficiency curves and—with time—damage to the impeller by pitting and erosion. Since all of these signs are obviously inexact, it became necessary to agree to apply certain ground rules so as to establish some uniformity for detecting cavitation.

The minimum (NPSH) is determined by a test in which both total head and efficiency are measured at a given speed and capacity under varying (NPSH) conditions. The results of such a test appear in a form similar to that in Fig. 20. At the higher values of (NPSH), head and efficiency remain substantially constant. As (NPSH) is reduced, a point is finally reached where the curves break, showing the impairment of pump performance caused by cavitation. The exact value of (NPSH) where cavitation starts is difficult to pinpoint. Usually, a drop of 3% in the head developed is taken as evidence that cavitation is occurring. For that particular speed and the capacity being tested, the (NPSH) that produces a 3% drop in head is stated to be the minimum required (NPSH).

(NPSH) tests of centrifugal pumps are normally carried out with cold water. Both the Hydraulic Institute Standards curves and pump manufacturers' rating curves indicate (NPSH) requirements for cold water. Thus, it might be assumed that (NPSH) required by a centrifugal pump for satisfactory operation is independent of the liquid vapor pressure at the pumping temperature. Actually this is not true.

Laboratory and field tests run on pumps handling a wide variety of liquids, and over a range of temperatures, have always shown that (NPSH) required for a given capacity and with a given pump apparently varies appreciably. For example, the required (NPSH) when handling some hydrocarbons is frequently much less than that required when the pump handles cold water. Even when pumping water, there is definite evidence that required (NPSH) decreases when the water temperature increases.

Altogether it became evident that the reduction in required (NPSH) must be a function of the vapor pressure and of the characteristics of the liquid handled by the pump. Thus, it was felt that rules could be developed to predict the effect of liquid characteristics on required (NPSH).

Such rules have been developed by the members of the Hydraulic Institute and incorporated in its Standards. We shall examine these rules, but before doing this, let us consider the effect of temperature on required (NPSH) for water, as this may help us better understand the effect of other liquids.

Performance with water

It has been noted that pumps handling hot water seem to require a lower (NPSH) than shown by cold-

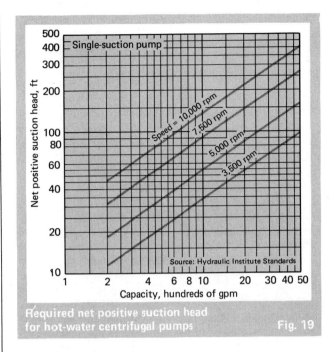

Required net positive suction head for hot-water centrifugal pumps — Fig. 19

water tests. The theory underlying this effect is fairly simple but need not be discussed in detail here. It is based on the fact that mild and partial cavitation can take place in a pump without causing extremely unfavorable effects.

The degree of interference with the proper operation of the pump caused by minor cavitation will bear a definite relationship to the temperature of the liquid handled by the pump. When we say that a pump is cavitating, we mean that somewhere within the confines of the pump, the pressure will have fallen below the vapor pressure of the liquid at the prevailing temperature. Thus, a small portion of the liquid handled by the pump will vaporize, and this vapor will occupy considerably more space within the impeller than did the equivalent mass of liquid.

If the pump is handling water at normal temperatures, the volume of a bubble of steam is tremendously larger than the volume of the original quantity of the

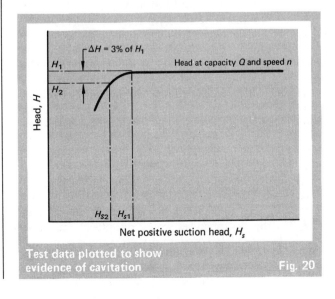

Test data plotted to show evidence of cavitation — Fig. 20

62 CENTRIFUGAL PUMPS

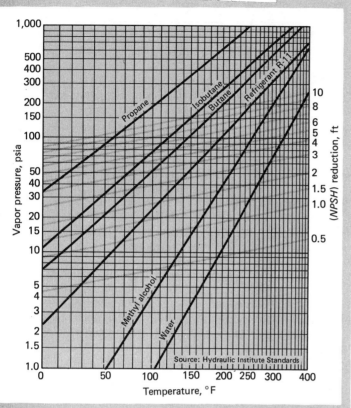

Reduction in net positive suction head for pumps handling hydrocarbon liquids   Fig. 21

water. For instance, at 50°F, one pound of water occupies 0.016 ft^3, while steam at the same temperature occupies 2,441 ft^3. The ratio of the two volumes is 152,500. This ratio diminishes as water temperature increases. At 212°F, one pound of water occupies 0.0167 ft^3, and one pound of steam, 26.81 ft^3, so that the ratio of volumes is only 1,605—almost one hundred times less than at 50°F. Thus, the higher the temperature of the water, the greater the reduction in (NPSH), that can be permitted for the same degree of effect on pump performance.

Performance with hydrocarbons

Pump applications for hydrocarbon processes frequently impose restrictive limitations on available (NPSH). On the other hand, it was found in the past that variations between the required (NPSH) when handling hydrocarbons and when handling cold water were generally favorable. These circumstances led both pump designers and designers of refineries to direct their efforts to understanding the phenomena and establishing rules that could be applied to predict the effect of any special liquid characteristics on the required (NPSH) of any centrifugal pump.

At first, it was thought that these variations did not exist and that if true vapor pressures or "bubble-point" pressures were to be used in the calculations of test (NPSH), the discrepancies would disappear and there would be complete correlation with water-test cavitation data. Corrections for (NPSH) with hydrocarbons were nevertheless used, as a matter of policy rather than being based on accepted theories. It was believed that a reduced (NPSH) for service with hydrocarbon fluids could be justified for two reasons:

1. Oil companies' specifications generally called for a maximum capacity and head at minimum (NPSH). In practice, it was unlikely that these two requirements would be imposed simultaneously. In fact, some field conditions are self-regulating—for instance, low capacity occurs at low (NPSH), as a result of a reduced flow in the system. Under these conditions, even if pump capacity falls off, (NPSH) is increased and equilibrium is eventually attained.

2. Cavitation with hydrocarbons was not as severe as with water, i.e., the head-capacity curve does not break off suddenly because (a) only the lighter fractions will boil first, and (b) the specific volume of hydrocarbon vapors is very small in comparison with that of water vapor.

Obviously, this does not tell the whole story. Many other factors affect the behavior of a pump handling hydrocarbons at low (NPSH). Thus, attempts to arrive at a more reasoned understanding continued, and led to a conversion chart for hydrocarbons by the Hydraulic Institute (since updated). The latest correction chart is incorporated in the 1975 edition of the Hydraulic Institute Standards [1] (see Fig. 21). To use this chart, enter with the pumping temperature and proceed vertically upward to the vapor pressure. From this point, follow along or parallel to the sloping lines to the right side of the chart, where reduction in (NPSH) may be read. If this value is greater than one-half of (NPSH) required with cold water, deduct one-half of the cold-water (NPSH) to obtain corrected required (NPSH). If the value read on the chart is less than one-half of cold-water (NPSH), deduct this value from the cold-water (NPSH) to obtain corrected required (NPSH).

Because of the absence of available data demonstrating (NPSH) reductions greater than 10 ft, the chart has been limited to that extent. Extrapolation beyond that limit is not recommended.

Warnings are included in the Hydraulic Institute Standards regarding the effect of entrained air or gases. This can cause serious deterioration of the head-capacity curve, of the efficiency, and of the suction capabilities, even when relatively small percentages of air or gas are present.

An exhaustive analysis of the phenomena that take place in a pump handling hydrocarbons is beyond the scope of our discussion. Such an analysis would at best be open to argument, because several somewhat conflicting interpretations still exist with respect to what actually takes place.

It is probably best to use the correction factor for the reduction in (NPSH) as an additional safety factor rather than as a license to reduce available (NPSH). This is a personal opinion, but one that is shared with a number of rotating-machinery specialists of some of the major petroleum and petro-chemical companies.

Inadequate suction conditions

When a system offers insufficient available (NPSH) for an optimum pump selection, there are several ways to deal with the problem. We can either find means to increase available (NPSH), or means to reduce required (NPSH), or do both.

To increase available (NPSH), we can:
1. Raise the liquid level.
2. Lower the pump.
3. Reduce the friction losses in the suction piping.
4. Use a booster pump.
5. Subcool the liquid.

To reduce required (NPSH), we can use:
6. Slower speeds.
7. A double-suction impeller.
8. A larger impeller-eye area.
9. An oversize pump.
10. Inducers ahead of conventional impellers.
11. Several smaller pumps in parallel.

Each of these methods presents advantages and disadvantages. We shall examine and evaluate these methods individually:

1. *Raise the liquid level*—At first glance, this appears to be the simplest solution unless it is impractical because (a) the liquid level is fixed, as in the case of a river, a pond or a lake; (b) the amount by which the level must be raised is completely impractical; or (c) the cost of raising a tank or a fractionating tower is excessive. Frequently, it will be found that only a few extra feet may permit the selection of a less-expensive or more-efficient pump, and the savings in first cost, energy or maintenance will far outweigh the additional costs incurred.

2. *Lower the pump*—Just as in the case of raising the liquid level, the cost of lowering the pump may not be as prohibitive as one might imagine because it may permit the selection of a higher-speed, less-costly and more-efficient pump. An alternative approach may be to use a vertical pump with the impeller located below ground level.

The penalty for this solution is that the pump bearings may have to be lubricated by the liquid being pumped. While successful bearing designs and materials have been developed for this purpose, it should be understood that the pump life cannot compare with the life obtainable from external bearings that are either grease or oil lubricated. Thus, one should expect more-frequent scheduled overhauls with this method.

3. *Reduce piping friction losses*—This is recommended under any circumstances; the cost of doing so will be easily repaid both by improved suction conditions and by savings in energy.

4. *Use a booster pump*—This solution is particularly effective in the case of pumps intended for high-pressure service, where the resulting permissible higher speeds will yield great savings in first costs of the main pump, higher efficiencies, and frequently a lesser number of stages—which in itself leads to greater reliability. The booster pump can be selected as a low-speed, low-head pump of single-stage design.

5. *Subcool the liquid*—This approach increases available (NPSH) by reducing the vapor pressure of the liquid being pumped. It is most readily accomplished by injecting liquid taken from somewhere in the stream where it is available at a colder temperature. In many cases, particularly at higher pumping temperatures, the amount of injected colder liquid is very small. As an example, if we are pumping water at 325°F, the injec-

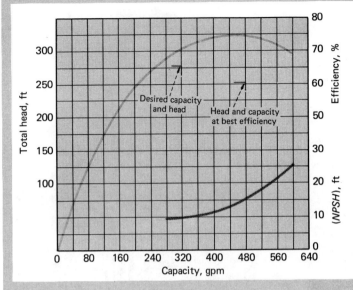

Effect of oversizing a pump   Fig. 22

tion of only 4% of 175°F water will subcool our stream to the point that available (NPSH) will have been increased by 20 ft.

6. *Use slower speeds*—Once a reasonable value of suction specific speed has been selected, it becomes obvious that the lower the pump speed, the lower the required (NPSH). The problem is that a lower-speed pump will be more expensive and less efficient than a higher-speed one selected for the same service. Thus, lowering the pump speed will seldom prove most economical.

7. *Use a double-suction impeller*—Particularly for larger capacities, whenever a double-suction impeller is available for the desired conditions of service, this presents the most desirable solution. It is based on the following:

If we select the same S value for both single- and double-suction impellers such that:

$$S = \frac{n_1 (Q_1)^{1/2}}{(H_{sr1})^{3/4}} = \frac{n_2 (Q_2)^{1/2}}{(H_{sr2})^{3/4}} \qquad (7)$$

where Subscript 1 refers to a single-suction impeller and Subscript 2 refers to a double-suction impeller.

Since $Q_2 = Q_1/2$, we can assume first that:

$$n_2 = n_1 \qquad (8)$$

in which case, $H_{sr2} = 0.63 \, H_{sr1}$, or:

$$H_{sr2} = H_{sr1} \qquad (9)$$

in which case, $n_2 = 1.414 n_1$.

By keeping the pump speed the same in both cases, as in Eq. (8), we can reduce $(NPSH)_R$ by 27% if we use a double-suction impeller. Alternatively, with a given $(NPSH)_R$, as shown in Eq. (9), we can operate a double-suction pump at 41.4% higher speed.

8. *Use a larger impeller-eye area*—This solution reduces required (NPSH) by reducing the entrance velocities into the impeller. These lower velocities may have little effect on pump performance at or near the pump's best efficiency point. But when such pumps run at partial capacity, this practice can lead to noisy operation, hy-

64 CENTRIFUGAL PUMPS

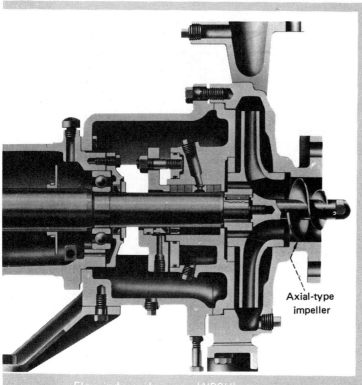

Flow inducer decreases (NPSH) requirements of a centrifugal pump   Fig. 23

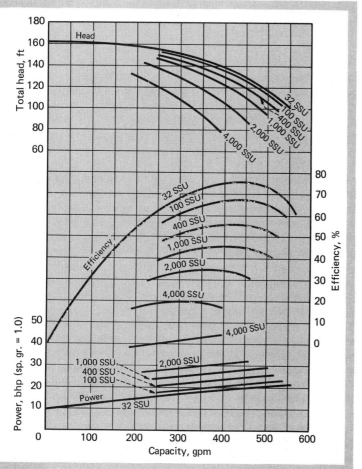

Liquid viscosity affects performance of centrifugal pumps   Fig. 24

draulic surges and premature wear. This problem will be discussed in greater detail later in the article. At this point, suffice to say that this is a dangerous procedure and should be avoided if possible.

9. *Use an oversized pump*—Because (NPSH) required by a pump decreases as the capacity is decreased, a larger pump than would otherwise be applied to the service is occasionally selected. This practice is risky and can lead to undesirable results. At best, the penalty is the choice of a more expensive pump that operates at a lower efficiency than might otherwise have been obtained (see Fig. 22). At worst, the operation at a lower percentage of the best efficiency flow can lead to exactly the same problems as the use of excessively enlarged impeller-eye areas.

10. *Use an inducer*—An inducer is a low-head axial-type impeller with few blades, which is located in front of a conventional impeller (Fig. 23). By design, it requires considerably less (NPSH) than a conventional impeller, so it can be used to reduce (NPSH) requirements of a pump, or to let it operate at higher speeds with given available (NPSH).

The inducer is an adequate answer for many situations but must be applied with care, as the permissible operating range of pumps with inducers is generally narrower than with conventional impellers.

11. *Use several smaller pumps in parallel*—Obviously, smaller-capacity pumps require lower (NPSH) values. While this appears to be a costly solution, it is not necessarily so. In many cases, three half-capacity pumps, of which one is a spare, are no more expensive than one full-capacity pump plus its spare. As a matter of fact, in many cases, two half-capacity pumps may be installed without a spare, since part-load can still be carried if one pump is temporarily out of service. In addition, if the demand varies widely, operating a single pump during light-load conditions will conserve energy, as we shall see later.

Viscosity and entrained gases

Earlier, we stated that pump performance is independent of the characteristics of the liquid being pumped, but qualified this by adding that the liquid viscosity does affect performance. This is because two of the major losses in a centrifugal pump are caused by fluid friction and disk friction. These losses vary with the viscosity of the liquid being pumped, so that both the head-capacity output and the mechanical output differ from the values they have when the pump handles water.

The performance of a pump tested first on water (viscosity = 32 Saybolt seconds universal, or SSU), and then on a variety of liquids having viscosities ranging from 100 to 4,000 SSU, is shown in Fig. 24. It is evident that by the time the viscosity reaches 2,000 SSU, pump performance will have deteriorated to such an extent that a positive-displacement pump will be more economical for the application.

It is not necessary to present here a complete discussion on the effect of viscosity on the flow of liquids. However, all correction factors for viscosity effects on pump performance have been developed experimen-

CENTRIFUGAL PUMPS AND SYSTEM HYDRAULICS

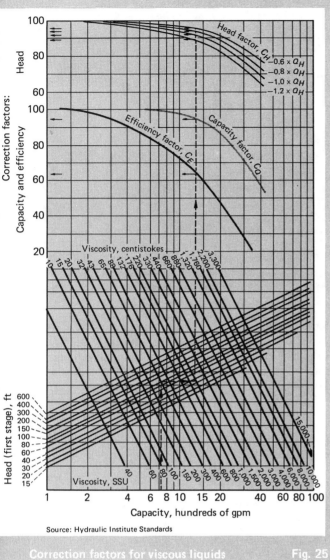

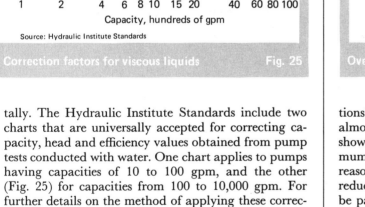

Correction factors for viscous liquids Fig. 25

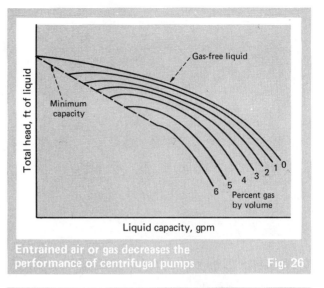

Entrained air or gas decreases the performance of centrifugal pumps Fig. 26

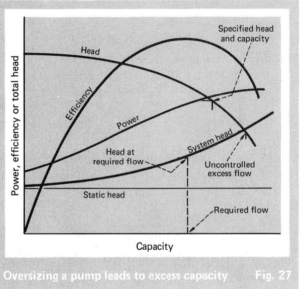

Oversizing a pump leads to excess capacity Fig. 27

tally. The Hydraulic Institute Standards include two charts that are universally accepted for correcting capacity, head and efficiency values obtained from pump tests conducted with water. One chart applies to pumps having capacities of 10 to 100 gpm, and the other (Fig. 25) for capacities from 100 to 10,000 gpm. For further details on the method of applying these correction factors to tests with water, and on the selection of a pump for given performance conditions at a given viscosity, consult the Standards [1].

Entrained air or gas

If entrained air or gas is permitted to enter a centrifugal pump along with the liquid, the performance of the pump will be unfavorably affected. The most frequent way that air enters the pump suction is by vortex formation at the free surface of the liquid. On occasion, air leaks into the pump through the pump stuffing box if proper precaution has not been taken to seal it properly. The amount of air or gas that can be handled with impunity by a centrifugal pump is probably in the range of ½% by volume (measured under suction condi-

tions). If the amount is increased to 6%, the effect is almost disastrous, as can be seen from a typical curve shown in Fig. 26. The dotted line indicates the minimum capacity at which the pump can be operated. The reason for this minimum is that if the pump capacity is reduced further than indicated, air or gas can no longer be partially swept out through the discharge, and the pump becomes air-bound.

Operation at off-design conditions

Theoretically, as long as available (NPSH) is greater than required (NPSH), a centrifugal pump is capable of operating over a wide range of capacities. As explained earlier, exact operating capacity is determined from the intersection of the pump's head-capacity curve with the system-head curve. This operating capacity can only be changed by altering either one or both of these curves—varying the pump speed will alter the head-capacity curve, while throttling the pump discharge will alter the system-head curve.

At any given speed, the performance of a centrifugal

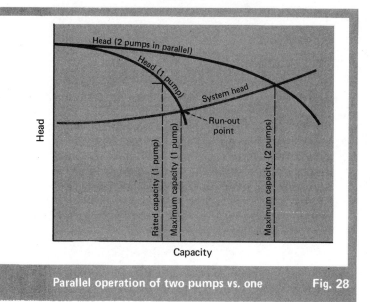

Parallel operation of two pumps vs. one   Fig. 28

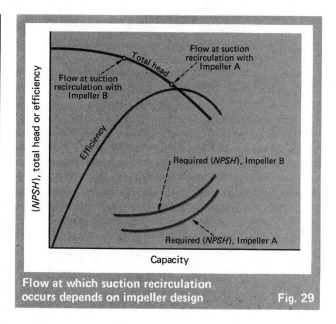

Flow at which suction recirculation occurs depends on impeller design   Fig. 29

pump is at its optimum at only one capacity point, i.e., the capacity at which the efficiency curve reaches its maximum. At all other flows, the geometric configuration of the impeller and of the casing no longer provides an ideal flow pattern. Therefore, our definition of "off-design" conditions must be any conditions wherein a pump is required to deliver flows either in excess or below the capacity at the best efficiency.

Operation at high flows

There are two circumstances that might lead to the operation of a pump at flows in excess of its best efficiency or even of its design point.

The first of these occurs when a pump has been oversized through the use of excessive margins in specifying both head and capacity. Under these circumstances, the pump performance and its relationship to the system-head curve might look as in Fig. 27. The head-capacity curve intersects the system-head curve at a capacity much in excess of the required flow, using excess power. Of course, the pump can be throttled back to the required capacity and the power reduced somewhat. But if, as frequently is the case, the pump runs uncontrolled, it will always run at the excess flow indicated in Fig. 27. Unless sufficient (NPSH) has been made available, the pump may suffer cavitation damage and, of course, power consumption will be excessive.

The second occurs when two or more pumps are used in parallel and one of them is taken out of service because demand has been decreased. Fig. 28 describes the operation of two such pumps. Whenever a single pump is running, its head-capacity curve intersects the system-head curve at flows in excess of design capacity. This is called the "run-out" point. Here, again, available (NPSH) and the size of the driver must be selected in such a manner as to satisfy conditions prevailing at this run-out point.

Operation at low flows

The most frequent cause for running a pump at reduced flows is the reduction in demand by the process served by the pump. However, it can also happen that two pumps operating in parallel may be unsuitable for this service at reduced flows, and one of the pumps on the line may have its check valve closed by the higher pressure developed by the stronger pump.

Operation of centrifugal pumps at reduced capacities leads to a number of unfavorable results that may take place separately or simultaneously, and must be anticipated or circumvented. Some of these are:

■ Operating at less than best efficiency—When reduced flows are required by the characteristics of the process, they can be handled by using a variable-speed drive, or by using several pumps for the total required capacity and shutting down the pumps sequentially as total demand is reduced. This last procedure will save energy, as will be described later on in this report.

■ Higher bearing-load—If the pump is of the single-volute design, it will be subjected to a higher radial thrust that will increase the load on its radial bearings. A pump that is expected to operate at such flows must be able to accept this higher bearing-load.

■ Temperature rise—as capacity is reduced, the temperature rise of the pumped liquid increases. To avoid exceeding permissible limits, a minimum-flow bypass must be provided. This bypass, which can be made automatic, will also protect against the accidental closing of the check valve while the pump is running.

■ Internal recirculation—At certain flows below that at best efficiency, all centrifugal pumps are subject to internal recirculation, in both the suction and the discharge area of the impeller. This can cause hydraulic surging and damage to the impeller metal similar to that caused by classic cavitation, but taking place in a different area of the impeller.

Internal recirculation

The first three unfavorable effects of operation at low flows are well understood and need no further explanation. On the other hand, the subject of internal recirculation has been little understood outside of a small number of pump designers. Because internal recircula-

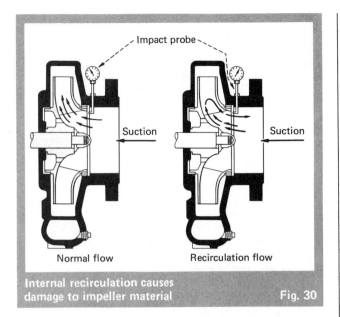

Internal recirculation causes damage to impeller material — Fig. 30

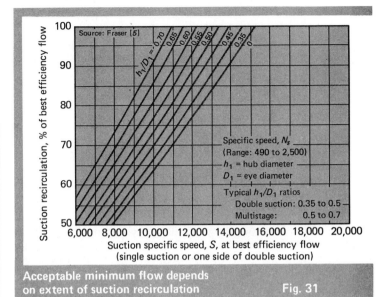

Acceptable minimum flow depends on extent of suction recirculation — Fig. 31

tion at the suction is most frequently the cause of field problems, we shall now discuss it.

The exact flow at which suction recirculation takes place depends on impeller design. The larger the impeller eye diameter and the larger the area at the impeller suction relative to its overall geometry (therefore, the lower the required (NPSH) at a given capacity and speed), the higher the capacity at which recirculation takes place as a percentage of the capacity at best efficiency (see Fig. 29).

Internal recirculation causes the formation of very intense vortices (Fig. 30) with high velocities at their core and, consequently, a significant lowering of the static pressure at that location. In turn, this leads to intense cavitation accompanied by severe pressure pulsations and noise, and this can be damaging to the operation of the pump and ultimately to the integrity of the impeller material.

Location of the material damage is an excellent way to identify whether the cause is classic cavitation or internal suction recirculation. If the damage is on the visible side of the inlet impeller vanes, the cause is classic cavitation. If the damage is to the hidden side (pressure side) of the vanes and must be seen with the help of a small mirror, the cause is suction recirculation.

Field problems caused by this phenomenon obviously occurred from time to time. But in the early 1960s, they became intensified. Two factors led to this greater incidence of trouble:

1. More pressure was exerted by the users to have pump designers reduce required (NPSH) values. This could only be achieved by increasing the impeller suction-eye areas, bringing the onset of internal recirculation closer to the best-efficiency capacity.

2. Higher heads per stage, and larger pump capacities, were increasing the energy levels of individual impellers, intensifying the unfavorable effects of internal recirculation.

Information on the phenomenon was first released in 1972 in a limited-circulation paper and later, more widely, in an article [3]. For obvious reasons, the mathematical solution was considered proprietary and was not published until 1981 [4,5].

Interim guidelines, however, were indicated suggesting that suction specific-speed values (S) not be allowed to exceed a range of 8,500 to 9,500 so as to avoid field problems if operation at significantly reduced flows were contemplated.

Fraser [4] gave exact formulas for calculating the flow at which internal recirculation would start at the suction, once certain geometric data were known about an impeller. He provided some close-approximation curves (of which Fig. 31 is one) in the event that these data are not readily available [5].

There is a way to confirm the value of the suction recirculation flow. An impact-head probe is installed into the direction of flow, as indicated in Fig. 30, with the probe directed into the eye. Under normal flow, the reading will be essentially the suction pressure less the velocity head at the eye. As soon as internal recirculation occurs, a flow reversal will show as a sudden rise in pressure.

The suction specific speed, S, of the pump must always be calculated for the conditions corresponding to the capacity at best efficiency. Guaranteed conditions of service may or may not correspond to this best-efficiency flow—they seldom do. S must also be calculated on the basis of pump performance with the maximum impeller diameter for which the pump is designed.

This constraint becomes obvious when we consider that internal recirculation at the suction occurs because of conditions that arise in and around the impeller inlet. These conditions are not necessarily affected by cutting down (i.e., reducing) the impeller diameter. Cutting down will move the best efficiency point to a lower flow value, but will not reduce the flow at which suction recirculation will occur.

Liquid characteristics do not affect the flow at which internal recirculation takes place. But they have a profound effect on the severity of the symptoms and on the extent of the damage. The reason is exactly the same as in the case of required (NPSH) to prevent the symptoms and damage caused by classical cavitation, as described earlier in this article. Thus, internal suction recirculation will always be less damaging when the pump handles hot water, and particularly when it handles hydrocarbons, than when it handles cold water.

Now that means are available to calculate the onset of internal recirculation, users and designers of centrifugal pumps should be in the position to establish sensible limits for minimum operating flows. To do so, it will be necessary to establish some guidelines between minimum flows and recirculation flows. But there's the rub! Setting up such guidelines is a fairly complex task for a number of reasons.

Operating a pump at flows below the recirculation flow leads to a variety of events—leading to unfavorable effects on pump performance and on the ultimate life of the impeller. We can lump all these events under the term "distress." In turn, the degree of distress will depend on a variety of factors, such as:
- Size of the pump—i.e. capacity, total head and horsepower.
- Value of the suction specific speed.
- Fluid characteristics.
- Materials of construction.
- Length of time the pump operates below certain critical flows.
- The degree of tolerance the pump user permits to the symptoms of distress exhibited by the pump.

The last factor makes the choice of minimum-flow guidelines a subjective one. Some users will unquestionably accept the fact that an impeller may have to be replaced every year; others will complain if an impeller on exactly the same service lasts only three or four years. Similarly, noise and pulsation levels perfectly acceptable to one user are cause for complaints by other users.

One observation may give some guidance to users on what constitutes an acceptable minimum flow with respect to the effect of the S value. Referring to Fig. 31 and assuming a hub-to-eye diameter, h_1/D_1, ratio of 0.45, an impeller having an S value of 14,000 will have its suction recirculation occur at about 100% of its best efficiency flow. If dealing with a fairly substantial pump handling cold water, one should be most reluctant to consider running it below this 100%. On the other hand, a pump with the same hub-to-eye-diameter ratio and an S value of 8,000 will have a recirculation flow of only 56% of best efficiency flow. I would have no hesitation running this pump when necessary at as little as 25% of best efficiency flow.

Vague as they may appear to be, here are some guidelines:
1. Unless there is a compelling reason to do so, do not specify ($NPSH$) values that result in S values much above 9,000.
2. In the case of relatively small pumps—say under 100 hp—the effect of suction recirculation is not apt to be as significant as for larger pumps.
3. Pumps handling hydrocarbons can be operated at lower flows than similar pumps handling cold water.
4. The risks of operating at flows much below the recirculation flow can best be determined after the pump is in operation. Provision should, therefore, be made to increase the minimum-flow bypass if there is a suspicion that too optimistic a decision as to this minimum flow was made at the time the pump was selected.
5. When the pump is not expected to operate at flows below its design condition [such as cooling-tower pumps operating in parallel, at constant speed and unthrottled (see Fig. 28)], higher S values can be used without concern for unfavorable effects from internal suction recirculation.

Energy conservation and pumps

The high cost of energy and the scarcity of fuels have become hard facts of life. More than ever, it is imperative to examine pump selection and operation with a view to minimizing power consumption. We might address ourselves to the pump designers and ask them what further improvements they may make in pump efficiencies. Alas, the easy improvements have already been incorporated in today's centrifugal pumps. The curves in Fig. 9 show the maximum efficiencies attainable from commercial pumps of different sizes and at different specific speeds. They have not changed materially since 1955.

Effect of specific speed

The higher the specific speed selected for a given set of operating conditions, the higher the pump efficiency and, therefore, the lower the power consumption. Barring other considerations, the tendency should be to favor higher specific-speed selections from the point of view of energy conservation.

Let us consider some typical examples (see Table II and III) wherein several alternative selections are examined for two sets of conditions. In both cases, it has been assumed that the design point corresponds to the best efficiency point of the pump.

In the first case (Table II), the difference between selections (2) and (3) does not appear to be significant enough to warrant considering a three-stage pump. On the other hand, a two-stage pump saves 20.5 hp, which at an energy cost of $300 to 400/(hp)(yr) will produce annual savings of $6,150 to $8,200.

Against these savings, we must consider a number of counterweighing factors. Among these: (1) the initial cost of a 2-stage pump will be higher than that of a single-stage pump, and (2) a two-stage design will preclude the use of a simpler overhung end-suction pump with a single stuffing box.

Similar factors must be weighed against the savings of 7.4 hp, or $2,220 to 2,960/yr, which favors a two-stage pump in the second case (Table III).

It may well be that the higher-specific-speed pumps will show sufficient savings to justify their selection. But are these savings really there? The answer to this question will depend entirely on the expected operating-capacity range of these pumps. The shape of the power consumption curve varies considerably with the specific speed of the pumps, as illustrated in Fig. 12. Before we can decide which of the possible selections is best in terms of energy consumption, we must examine power consumption not only at the design point but also over the entire range of capacities that the pump will encounter in service.

Let us expand our analysis of the case examined in Table III. In addition to the power at the 100% design capacity, we shall compare the power consumption for single- and two-stage pumps at 75% and 50% flow. (see Table IV). Much to our surprise, we see that instead of saving energy at all flows, the two-stage pump uses 7.6 hp more at 75% flow, and 18 hp more at 50%.

To establish the real energy balance between the two selections, we need to predict the subdivision of operating hours at various loads. If we assume that this subdivision will correspond to that shown in Table V, we find that the most efficient pump is not the best selection and that the single-stage pump will save 36,267 hp-h yearly over the two-stage pump. Obviously, the final answer will always depend on the load factor for a given installation.

Once the optimum combination of specific speed and (NPSH) requirements has been established for any given set of operating conditions, we still must choose between various pump offerings that may differ somewhat in guaranteed efficiencies.

Our most obvious reaction might be to look with greater favor on pumps having higher efficiencies, favoring those that might exceed others by as little as ½% or 1%. All things being equal, there is some logic to this approach. But, all things are not always equal and small differences in guaranteed efficiencies may have been obtained at the expense of reliability, by using smaller running clearances or lighter shafts.

Savings in power consumption obtained from these small differences in efficiency are insignificant when we compare them with other approaches for reducing or eliminating wasteful power consumption. While we cannot examine every approach available, there are three particular sources of significant power savings that I wish to analyze in some detail. These are (a) savings of power wasted by oversized pumps, (b) savings that may be obtained by operating the minimum number of pumps in multiple-pump installations, and (c) savings that can be obtained by restoring internal clearances at the right time.

Effects of oversizing

One of the greatest sources of power waste is the practice of oversizing a pump by selecting design conditions having excessive margins in both capacity and total head. It is strange, on occasion, to see a great deal of attention being paid to a one-point difference in efficiency between two pumps, while at the same time potential savings of 5, 10 or even 15% of the power are ignored through an over-conservative attitude in selecting the required conditions of service.

Still, it is true that some margin should always be included, mainly to provide against the wear of internal clearances that with time will reduce the effective pump capacity. How much margin to provide is a fairly complex question, because the wear that will take place varies with the type of pump, the liquid handled, the severity of the service, and a number of other variables. We shall examine this question later when we analyze the savings in power consumption that can be realized from restoring internal clearances to their original values.

A centrifugal pump operating in a given system will deliver a capacity corresponding to the intersection of its head-capacity curve with the system-head curve—provided that available (NPSH) is equal to or exceeds required (NPSH). To change this operating point requires changing either the head-capacity curve or the system-head curve or both. The first can be accomplished by varying the speed of the pump or changing the impeller diameter, while the second requires altering the friction losses by throttling a valve in the pump discharge.

In the majority of pump installations, the driver is a

Pump selections for 2,000 gpm, 700-ft head — Table II

Selection	Stages, No.	Speed, rpm	Specific speed, N_s	Chart efficiency, %	Power, bhp
1	1	3,550	1,167	81	436.5
2	2	3,550	1,962	85	416
3	3	3,550	2,659	85.5	413.4

Pump selections for 2,000 gpm, 400-ft head — Table III

Selection	Stages, No.	Speed, rpm	Specific speed, N_s	Chart efficiency, %	Power, bhp
1	1	3,550	1,775	83	243.4
2	2	3,550	2,985	85.5	236

Part-load power consumption for pumps (see Table III: 2,000 gpm, 400-ft head) — Table IV

Selection	Specific speed N_s	Design flow, %	Power, % of design bhp	Actual power, bhp	Gain or loss compared to selection
1	1,775	100	100	243.4	Base
1	1,775	75	89	216.6	Base
1	1,775	50	76	185	Base
2	2,985	100	100	236	7.4 hp gain
2	2,985	75	95	224.2	(7.6 hp loss)
2	2,985	50	86	203	(18 hp loss)

Comparison of yearly operation at various loads — Table V
(One- and two-stage pumps at 2,000 gpm, 400-ft head)

Capacity, % of design flow	Operating time, %	h	Advantage, hp-h Single-stage	Two-stage
100	30	2,628	—	19,447
75	60	5,256	39,946	—
50	10	876	15,768	—
Total	100	8,760	55,714	19,447

Net saving in favor of single-stage pump = 36,267 hp-h

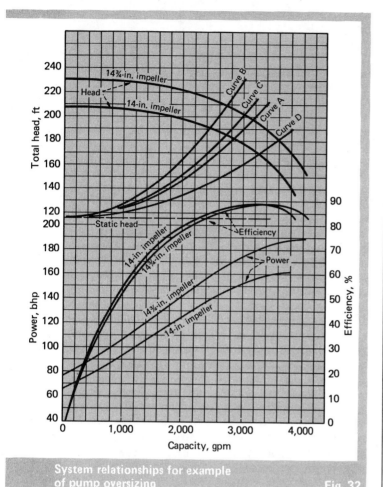

Fig. 32 System relationships for example of pump oversizing

constant-speed motor. This is the last means that is used to change the pump capacity. Thus, if we have provided too much excess margin in the selection of the pump head-capacity curve, the pump will have to operate with considerable throttling to limit its delivery to the desired value. On the other hand, if we permit the pump to operate unthrottled, which is more likely, flow into the system will increase until that capacity is reached where the system-head and head-capacity curves intersect.

The old "rules of thumb" on selecting pipe sizes and valve sizes should be questioned. We should evaluate carefully the economy of investing in larger pipes and valves having lower friction losses, against the long-term power savings resulting from these lower friction losses.

Example of oversizing

Let us examine a pumping system for which the maximum desired capacity is 2,700 gpm, static head is 115 ft, and total friction losses, assuming 15-yr-old pipe, are 60 ft. Total head required at 2,700 gpm is 175 ft.

We can now construct a system-head curve, shown as Curve A in Fig. 32. If we add a margin of about 10% to the desired capacity and, as is frequently done, add some margin to the total head above the system-head curve at this rated flow, we end up by selecting a pump for 3,000 gpm and 200 ft total head. The performance of such a pump, with a 14¾-in. impeller, is superimposed on Curve A.

This pump develops excess head at the maximum desired capacity of 2,700 gpm. If we wish to operate at that capacity, this excess head will have to be throttled, as shown by Curve B.

At 3,000 gpm, the pump will take 175 bhp and we will have to drive it with a 200-hp motor. At the desired capacity of 2,700 gpm, operating at the intersection of its head-capacity curve and Curve B, the pump will absorb 165 bhp.

Thus, the pump has been selected with too much margin. We can safely select a pump with a smaller impeller diameter—say 14 in. The head curve for the smaller impeller will intersect Curve A at 2,820 gpm, giving us about 4% margin in capacity, which is sufficient. We will still have to throttle the pump slightly, and our system-head curve will become Curve C. The power consumption at 2,700 gpm will now be only 145 bhp instead of the 165 bhp required with our overly conservative selection. This is a respectable 12% saving in power consumption. Furthermore, we no longer need a 200-hp motor; a 150-hp motor will do quite well. The saving in capital expenditures is another bonus from not oversizing.

Our savings may actually be even greater than we have shown. In many cases, the pump may be operated unthrottled—the capacity being permitted to run out to the intersection of the head-capacity curve and Curve A. If this is so, a pump with a 14¾-in.-dia. impeller would operate at approximately 3,150 gpm and take 177 bhp. If a 14-in. impeller were to be used, the pump would operate at 2,820 gpm and take 148 bhp. We could be saving over 16% in power consumption.

Our real margin of savings is actually greater than indicated. The friction losses we used to construct the system-head curve (Curve A) were based on losses through 15-yr-old piping. The losses through new piping are only 0.613 times the losses we have assumed. The system-head curve for new piping is shown as Curve D. If the pump we had originally selected (with a 14¾-in. impeller) were to operate unthrottled, it would run at 3,600 gpm and take 187.5 bhp. A pump with a 14-in. impeller would intersect the system-head curve D at 3,230 gpm and take 156.5 bhp, with a saving of almost 16.5%.

As a matter of fact, we could use a 13¾-in. impeller. Its head-capacity curve would intersect Curve D at 3,100 gpm and take 147 bhp, now saving 21.6%.

Important energy savings can be made if, at the time of selecting the conditions of service, reasonable restraints are exercised to avoid using excessive safety margins for obtaining the rated service conditions.

But what of existing installations in which the pump or pumps have excessive margins? Is it too late to achieve these savings? Far from it! It is possible to accurately establish the true system-head curve by running a performance test once the pump has been installed and operated. A reasonable margin can then be selected, and three options become available:

1. The existing impeller can be cut down to meet the conditions of service required for the installation.
2. A replacement impeller with the necessary re-

duced diameter may be ordered from the pump manufacturer. The original impeller is then stored for future use if friction losses are ultimately increased with time or if greater capacities are required.

3. In certain cases, there may be two separate impeller designs available for the same pump—one of which is of narrower width than the one originally furnished. A narrow replacement impeller may then be ordered from the pump manufacturer. Such a narrower impeller will have its best efficiency at a lower capacity than the normal-width impeller; it may or may not need to be of smaller diameter than the original impeller, depending on the degree to which excessive margin was originally provided. The original impeller is put away for possible future use.

Variable-speed operation

While the majority of motor-driven centrifugal pumps are operated at constant speed, some pumps take advantage of the possible savings in power consumption provided by variable-speed operation.

Wound-rotor motors were once frequently used for this purpose, but present practice generally interposes a variable-speed device such as a magnetic drive or a hydraulic coupling between the pump and the electric motor. Alternatively, we can use a variable-voltage, variable-frequency motor control that has the great advantage of maintaining essentially a constant motor efficiency, regardless of operating speed. On the other hand, a variable-speed device such as a hydraulic coupling has slip losses, so that its efficiency decreases directly as the ratio of output speed to input speed. Variable-frequency operation has an additional advantage: It permits an electric motor to operate above the synchronous speed as well as below it.

Variable-speed operation makes it possible to meet the required conditions of service without throttling, by reducing the pump operating speed. Variable-speed drives—especially by means of variable-frequency input—will be used more frequently in the future.

It is not necessarily too late to achieve power savings in some existing installations by converting them to variable-speed operation. To decide whether to make such a modification, it is necessary to plot the actual system-head curve, to calculate the speed required at various capacities over the operating range, and to determine the motor horsepower output over this range, including losses, if any, incurred in variable-speed devices. The difference between this horsepower and the pump brake horsepower at constant speed represents potential power savings at these capacities.

It is then necessary to assign a predicted number of hours of operation at various capacities, and to calculate the potential yearly savings in hp-h or kWh. These savings are translated into cost savings and can be used to determine whether the cost of the modification to variable-speed operation is justified.

Run one pump instead of two

Many installations are provided with so-called "half-capacity" pumps, with two pumps operating in parallel to deliver the required flow under full load. If the service on which these pumps are installed is such that the required flow varies over a considerable range, important power savings may be available through improved operating practices. Too often, both pumps are kept on the line even when the demand drops to a point where a single pump can carry the load (see Fig. 28). The amount of energy wasted in running two pumps at half-load when a single pump can meet this condition is significant. This can best be demonstrated by referring to a so-called 100% pump curve, as shown in Fig. 10a.

To simplify our analysis, let us neglect the question of capacity or pressure margins and imagine that the pumps carry their full load with throttling valves wide open. We shall also assume that these pumps are operating at constant speed. Thus, full-load conditions are met by running two half-capacity pumps, each operating at its 100% capacity point and each taking 100% of its rated brake horsepower.

If we want to reduce the flow to half load and still maintain both pumps on the line, it will be necessary to throttle the pump discharge and create a new system-head curve. Under these conditions, each pump will deliver 50% of its rated capacity at 117% rated head, much of which will have to be throttled. Each pump will take 72.5% of its rated power consumption. Thus, the total power consumption of two pumps operating under half-load conditions would be 145% of that required if a single pump were to be kept on line.

Instead, we could stop one of the pumps, and carry the half-load requirement, with a single pump running at 100% of rated capacity. The discharge would be throttled considerably less than if both pumps were left running. Power consumption would be 100% of the rated value for a single pump. Running a single pump when the process load is down to 50% results in a power saving of 31% over running both pumps.

As a matter of fact, a single pump could carry much more than the capacity corresponding to half-load, since the head-capacity curve of a single pump might intersect with the system-head curve at anywhere from 60 to 70% of load, depending on the exact shape of the system-head and the head-capacity curves.

Other benefits are to be had from such an operating practice. In the first place, if we assume 8,500 yearly operating hours, of which 20% take place at flows of 50% of maximum flow or lower, each pump will operate for only 7,650 h/yr instead of 8,500, extending the calendar life of all the running parts by over 11%.

Pumps that frequently operate at reduced capacities do not have as long a life as pumps operated at close to their best efficiency point. Thus, running only one pump whenever it can handle the required flow will add much more life to each pump than just the difference in operating hours.

Restoring internal clearances

The rate of wear of pump parts at internal clearances depends on many factors. To begin with, it increases in some relation to the differential pressure across the clearances. It also increases if the liquid pumped is corrosive or contains abrasive matter. On the other hand, the rate is slower if hard, wear-resisting materials are used for the parts subject to wear. Finally, wear can be

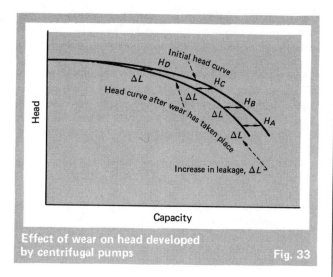

Effect of wear on head developed by centrifugal pumps   Fig. 33

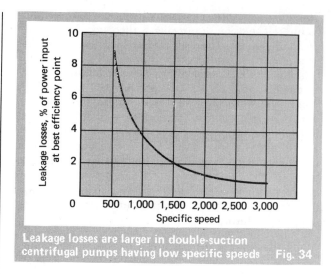

Leakage losses are larger in double-suction centrifugal pumps having low specific speeds   Fig. 34

accelerated if momentary contact between rotating and stationary parts occurs during pump operation.

When the running clearances increase with wear, a greater portion of the gross capacity of the pump is short-circuited through the clearances and has to be repumped. The effective or net capacity delivered by the pump against a given head is reduced by an amount equal to the increase in leakage.

While in theory the leakage varies approximately with the square root of the differential pressure across a running joint and, therefore, with the square root of the total head, it is sufficiently accurate to assume that the increase in leakage remains constant at all heads. Fig. 33 shows the effect of increased leakage on the shape of the head-capacity curve of a pump. Subtracting the additional internal leakage from the initial capacity at each head gives a new head-capacity curve after wear has taken place.

We must compare the cost of restoring the internal clearances after wear has taken place with the value of the power savings that we may obtain from operating a pump having the original clearances. This cost is relatively easy to determine. We can obtain prices on new parts and estimate the cost of labor required to carry out the task. But how about the savings?

These savings are not the same for every pump. Both analytical and experimental data have indicated that leakage losses vary with the specific speed of a pump.

Fig. 34 shows the relationship between the leakage losses of double-suction pumps and their specific speeds.

If, for instance, we are dealing with a pump having a specific speed of 2,500, the leakage loss in a new pump will be about 1%. Thus, when the internal clearances will have increased to the point that this leakage will have doubled, we can regain approximately 1% in power savings by restoring the pump clearances. But if we are dealing with a pump having a specific speed of 750, it will have leakage losses of about 5%. If the clearances are restored after the pump has worn to the point that its leakage losses have doubled, we can count on 5% power savings.

Thus, restoring the clearances of the lower-specific-speed type pumps yields greater returns in terms of the reduction of leakage losses. In addition, lower-specific-speed pumps generally have higher heads per stage than higher-specific-speed pumps. All else being equal, wear is increased with higher differential heads. Therefore, we will generally find more reasons to renew clearances of high-head pumps, and more savings.

References

1. "Hydraulic Institute Standards," 13th ed., Hydraulic Institute, Cleveland, Ohio, 1975.
2. Karassik, I. J., Krutzch, W. C., Fraser, W. H., and Messina, J. P., "Pump Handbook," McGraw-Hill, New York, 1976.
3. Bush, A. R., Fraser, W. H., and Karassik, I. J., Coping With Pump Progress: The Sources and Solutions of Centrifugal Pump Pulsations, Surges and Vibrations, *Pump World*, Worthington, Mountainside, N.J., Summer 1975 and March 1976.
4. Fraser, W. H., Recirculation in Centrifugal Pumps, Winter Annual Meeting of ASME, Nov. 16, 1981, ASME, New York.
5. Fraser, W. H., Flow Recirculation in Centrifugal Pumps, Turbomachinery Symposium, Texas A&M University, College Station, Tex., December 1981.

The author

Igor J. Karassik is chief consulting engineer, Worthington Div., McGraw-Edison Co., 233 Mount Airy Road, Basking Ridge, NJ 07920. He was appointed to the position in December 1976 after spending 42 years in various capacities at Worthington on the research, design and application of single- and multistage pumps. He has written more than 450 articles on pumps, and is the author or coauthor of four books and a handbook on pumps. He has a B.S. and M.S. from Carnegie Institute of Technology, is a Life Fellow of ASME, and a professional engineer in New Jersey. He is also a member of Tau Beta Pi, Pi Tau Sigma and Sigma Xi.

Unusual problems with centrifugal pumps

Solutions to these puzzling pump problems may help you to resolve some of the difficulties that you encounter with pumps.

S. Yedidiah,
Centrifugal Pump Consultant

At first sight, a centrifugal pump seems to be one of the simplest of machines. In practice, however, it is capable of posing an enormous spectrum of different problems [1]. Occasionally, one comes across problems that seem to defy everything we know about centrifugal pumps. We shall report on four such cases.

Case No. 1 — The twin pumps

In a chemical process plant, two identical pumps were installed side by side, to transfer liquid from the same source into the same pressurized container. Each pump had been provided with a separate suction and a separate discharge line. Also, the two pumps were never used simultaneously. While one ran, the other served as a standby.

Everything about these two systems seemed to be identical — except that one pump performed perfectly, whereas the second operated with great noise and vibration. The troublesome pumping system was dismantled several times, but nothing wrong could be found.

The author realized that the successful pump's pipe loop had a 2-in.-dia. discharge, with a reducer connecting directly to a 1½-in. pipeline. However, the troublesome loop had a 6-ft length of of 2-in.-dia. pipe connected to the pump's discharge, and only after this length was the line reduced to 1½ in.

When the 2-in.-dia. section of pipe was replaced by 1½-in. pipe, the problem-causing pump operated satisfactorily.

The reason can be explained by referring to Fig. 1. Up to a certain critical flowrate, Q_c, the net-positive-suction-head (NPSH) requirements of a centrifugal pump increase, approximately, as the square of the flowrate [2]. Above Q_c, however, the NPSH requirements start to increase at a much faster rate — shown schematically in Fig. 1.

In the particular case described above, the frictional losses in the pipelines constituted a very significant part of the total head against which each of these two pumps had to operate. In the discharge line that consisted exclusively of 1½-in.-dia. piping, the resistance to the flow was adequately high. This kept the total head — against which the pump had to operate — well above H_c, the critical pumping head. This, in turn, limited the flowrate to well below Q_c. In the other pipeline, however, the reduced resistance of the 2-in.-dia. pipe section, brought down the total head to well below H_c. This increased

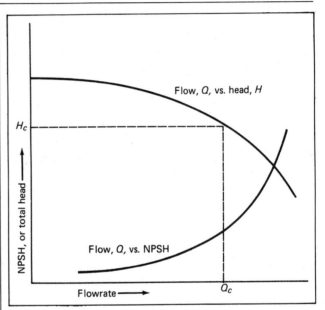

Figure 1 — Effect of flowrate on a pump's NPSH requirements

the NPSH requirements of the pump well above the available NPSH. Consequently, cavitation developed within the pump. This, in turn, gave rise to the noise and vibration.

Case No. 2 — Head in a storage tank

A somewhat related case was encountered with a pump installed in an oil depot. The pump transferred fuel from a storage tank to oil-delivery trucks. Whenever the oil level in the storage tank was low (i.e., when the available NPSH was low), the pump operated satisfactorily. However, when the storage tank became full (i.e., when the available NPSH was high), the pump operated with extreme noise and vibration.

The total head against which a pump operates is defined as the *difference* between the total head existing at a pump's outlet, and the total head available at its inlet. In this particular application, the discharge head tended to be practically constant. This meant that an increase in the available NPSH automatically reduced the total head against which the pump had to operate.

Thus, in accordance with what has been explained in connection with Fig. 1, the author concluded that when the storage-tank was full, the pump operated well below the

Originally published December 8, 1986

critical head, H_c. This, however, meant that the pump delivered a flowrate significantly higher than Q_c. However, the flowmeter that was installed in the pipeline, as well as measurements of the time required to fill a a fuel-truck's tank, indicated that the flowrate was well below Q_c. There seemed to be no solution to the puzzle.

However, the author recalled a study he had made earlier on the effects of flowrate on NPSH requirements [3]. According to that study, the most important factor that determines the NPSH requirements of a pump at a given speed is the rate of flow through the impeller.

The flowrate through an impeller is usually slightly greater than that through the pipeline, owing to leakage through the wearing rings. However, if a wearing ring is missing, this short-circuits the impeller discharge to the impeller eye. In such a case, the flow through the impeller may easily be 30 to 40% higher than the flow through the pipeline. In our particular case, this would have brought the total flow through the impeller well above the critical flowrate, Q_c.

The pump was opened, and the front wearing ring was found to be missing. A new wearing ring was installed in the casing, which solved the problem.

Case No. 3 — A recessed-hub impeller

A pump that had operated satisfactorily at the NPSH values presented in Fig. 2, Curve A, showed satisfactory performance at 40 gpm, when operating at an available NPSH of 5 ft. The same pump, however, continually failed at much lower flowrates (Curve A, dashed line), although previous tests (Curve A, solid line) showed that it could operate, at these flowrates, at NPSH-values significantly lower than the available 5 ft. Its performance at these times is shown by Curve B in Fig. 2 [4].

In order to eliminate this problem, the impeller was replaced by another of identical design. This time, the pump produced satisfactory results (Fig. 2 Curve C). But the question still remained: Why did the first impeller fail?

A close inspection of the two impellers revealed that the failing impeller had a recessed hub, as shown in Fig. 3 at A, while the second impeller had a solid hub (with no recess). (The same series of tests was repeated with a second, identical pump, with the same results.)

In order to verify whether this difference in the castings was the real cause of the observed differences in performance, the recess of one of the impellers was filled with epoxy and redrilled, as shown in Fig. 3 at B. This immediately restored the full suction capability of the pump, as seen by the test results presented in Fig. 2 by the dashed Curve D.

The tests were performed on pumps ordered by a customer, and there was no time to find out what caused the above effect. It seems, however, that the balancing holes acted here as resonators. The natural frequency of such resonators depends upon their length. Most probably, the shortening of the balancing holes brought their natural frequency in unison with the periodical shedding of vortices, which are known to appear at low partial flowrates. This, in turn, caused the early appearance of cavitation.

This problem occurred at a pump-manufacturer's plant [4], but such a case can easily occur in the field when a worn-out impeller is replaced by a new one. This is especially true when the spare part has not come from the original source.

Case No. 4 — The inconstant pump

This problem, too, was encountered at a pump factory [5], and is also one that can sometimes occur in the field.

A pump designated to operate at 50% of its best efficiency point, when provided with 3 ft of NPSH, was tested at a constant value of NPSH and found to operate perfectly. The pump test was then rerun in the presence of a witness. This time, however, the pump failed completely.

The pump was tested still another time, this time at the constant specified flowrate and variable NPSH. The test produced a two-level curve—Curve A of Fig. 4 [5].

In view of the unsatisfactory and confusing results ob-

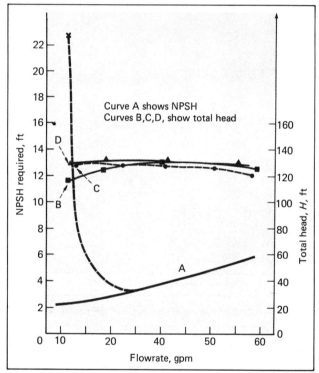

Figure 2 — Effect of length of balancing holes on the suction performance of a centrifugal pump

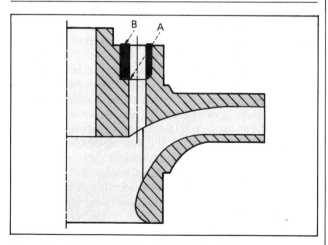

Figure 3 — Modification of impeller hub in tested pump

tained from the tested pump, both the installation and all test procedures were carefully scrutinized. However, there was no clue as to the source of the observed inconsistencies. As a final check, the entire loop was put under vacuum, and held for 20 min. The whole pumping system was proven to be adequately airtight.

As a last resort, it was decided to find if the problem could have been due to air dissolved in the pumped liquid.

A vacuum pump, located at the top of the airtight suction tank, was allowed to run and, with the aid of regulating valves, air removal from the tank was adjusted to such a rate as to keep a constant vacuum of 15 in. Hg.

After holding this vacuum for about 15 min, to allow a part of the dissolved air to escape, the pump was tested again.

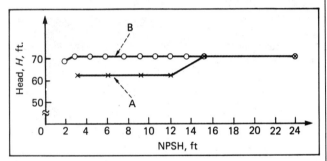

Figure 4 — Time effects of dissolved air on pump's NPSH

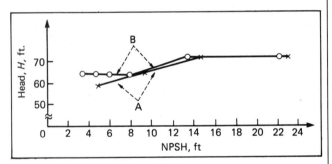

Figure 5 — How dissolved air caused still another failure

This test produced satisfactory results (Fig. 4, Curve B).

Using the same procedure, the test was repeated in the presence of a witness, and the pump failed again! This time, the head vs. NPSH curve had a shape as shown in Fig. 5 by Curve A. A series of additional tests were then carried out, but the results were never consistent. In one particular case, they even produced a curve such as that in Fig. 5, Curve B.

After a significant amount of testing, the answer to the mystery was finally found.

The observed inconsistencies in the test results turned out to be actually due to the air dissolved in the liquid. In addition, however, they were found also to depend upon the number of steps in which the NPSH was lowered to 3 ft, as well as upon the time lapse between two sets of readings.

The first (preliminary) test was carried out at a constant NPSH, and varying flowrate. The tank had been evacuated to 27 in. Hg, and kept under that vacuum until the test engineer was ready to make the readings. This took enough time to allow the vacuum pump to remove any excess of air that had been liberated from the water. Consequently, the pump performed as expected.

With regard to the tests at constant flowrate and varying NPSH, the time effect can be explained as follows:

The initial removal of air was carried out at a vacuum of 15 in. Hg. Consequently, below this pressure, additional air was being liberated from the liquid; this air appeared as a mass of tiny bubbles, dispersed throughout the volume of water. This lowered the specific weight of the pumped mixture.

The amount of air liberated each time the pressure had been lowered depended, of course, on the magnitude of the drop in pressure: The greater the reduction in pressure, the more air was liberated.

When the reduction in pressure was carried out in many small steps, as in the case of Fig. 4, Curve B, only a small amount of air was liberated with each step. This allowed enough time for the vacuum pump to remove the liberated air during the periods between the measurements. Therefore, the centrifugal pump produced satisfactory results.

When the measurements were carried out in large steps, as in Fig. 5 (Curve A), the rate at which the air was liberated from the water exceeded the rate at which it was removed by the vacuum pump. This resulted in a gradual increase in the amount of free air dispersed throughout the pumped liquid, and caused a continuous drop in pressure.

Sometimes, at an intermediate number of steps, a state of equilibrium occurred between the rate at which air was liberated from the liquid, and the rate of air removed by the vacuum pump. In such a case, the pressure readings remained constant, but were lower due to either a reduction in the specific gravity of the liquid caused by the presence of air bubbles, or partial air blockage of the impeller inlet, or both [1]. This produced a bilevel curve, like Curve A in Fig. 4.

In the case shown in Fig. 5 Curve B, the available NPSH was lowered, in large steps, down to about 8 ft. This caused the air to be liberated from the pumped water at a higher rate than that which was removed by the vacuum pump. This caused a significant drop in the total head. Below 8-ft NPSH, however, the mechanic started to reduce the available NPSH in significantly smaller steps. This enabled the pump to start to recover a part of the lost head.

References

1. Yedidiah, S., "Centrifugal Pump Problems, Causes and Cures," Penn-Well Book Co., Tulsa, Okla., 1980.
2. Yedidiah, S., Some Observations Relating to Suction Performance of Inducers and Pumps, *J. Basic Eng.*, Sept. 1972.
3. Yedidiah, S., Effect of Scale and Speed on Cavitation in Centrifugal Pumps, ASME Symposium on Fluid Mechanics in the Petroleum Industry, Dec. 1975.
4. Yedidiah, S., A Possible Explanation to Some Puzzling Cavitation Phenomena, ASME Cavitation and Polyphase Flow Forum, 1982.
5. Yedidiah, S., Time-Effects of Air on NPSH-Tests, ASME Cavitation and Multiphase Flow Forum, 1983.

The author

S. Yedidiah is a centrifugal pump consultant, 89 Oakridge Rd., West Orange, NJ 07052, Tel: (201) 731-6293. He has been an active pump specialist since 1938, and has worked for a number of pump manufacturers, including Worthington Pump Inc. He has published many articles, and is the author of "Centrifugal Pump Problems—Causes and Cures," published by Penn-Well Books. He is a member of the American Soc. of Mechanical Engineers, and is a registered professional engineer.

Startup of centrifugal pumps in flashing or cryogenic liquid service

Inder S. Rattan* and Vijay K. Pathak†

Startup of centrifugal pumps in nonflashing service usually is easy, basically requiring priming to remove gases and vapors from the pump casing. The pump is started with the discharge valve closed to prevent motor overload, then the valve is gradually opened fully to allow the pump to operate against system pressure. However, in liquid-hydrocarbon flashing and cryogenic service, pump startup is more difficult due to flashing of the liquid.

There are generally two operations during which flashing may occur in the pump:

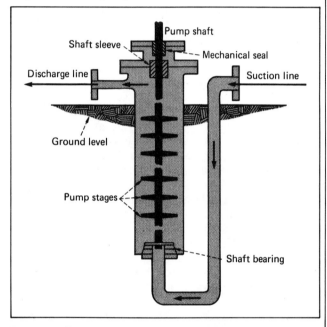

Figure 1 — Typical multi-stage vertical centrifugal pump

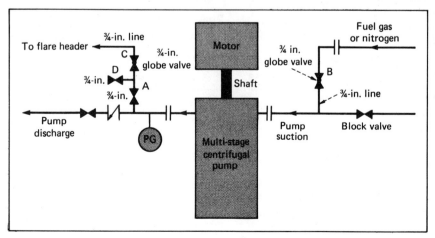

Figure 2 — Recommended arrangement for pump startup

1. During plant commissioning, or after a prolonged plant shutdown when the whole system has been depressured to atmospheric conditions. The pumps are not part of the pressurized system, as the unit is just being brought up to operating pressure.
2. During actual plant operation, when it may become necessary to bring a spare pump online. Although in some cold locations the spare is kept full of liquid, this is not the general practice (usually, the spare is kept purged and depressured) because of problems resulting from vaporization of the volatile liquid.

In some cases, a small liquid stream is bled through a hole in the check valve of the spare pump (or through a restriction orifice in a line bypassing the check valve) to renew the liquid in the unit and keep it cool. This would require block valves on the spare unit's suction and discharge lines to be open all the time, which may not please the plant operations people. Moreover, in very hot areas such as the Middle East, a fairly large amount of liquid might have to be bled back to keep the liquid in the spare pump cool, and this would have to be accounted for in the pump design. In addition, there is a possibility that keeping liquid in the spare pump may overpressure the system to above its design pressure.

Effects of flashing

The extent of flashing depends upon the fluid being pumped, but flashing has two basic effects. First, it creates a vapor-lock situation where vapor generation prevents head buildup and causes cavitation, and the pump cannot be started up. Second, in the case of light liquefied products (both hydrocarbon and nonhydrocarbon), flashing within the pump can create very low temperatures and cause auto-refrigeration.

In the case of liquid ethane, for example, temperatures as low as –60°F could be reached, causing the pump material and mechanical specifications to change significantly, since the unit is generally not designed for such low temperatures. Several installations with which we are familiar have experienced failure of such components as mechanical seals, thrust bearings, and wearing rings during pump startup. In addition to being costly in terms of repair, the downtime could result in significant losses.

Starting up the pump

The following method for starting up centrifugal pumps in flashing or cryo-

*Stone & Webster Canada Ltd., 2300 Younge St., Toronto, Ont. M4P 2W6, Canada.
† Monenco Engineers & Constructors, Inc., 801 6th Ave., S.W., Calgary, Alta. T2P 3W3, Canada.

FLASHING OR CRYOGENIC LIQUID SERVICE

genic service is of general applicability. Fig. 1 illustrates the main components of a typical multistage pump arrangement, and Fig. 2 shows the recommended setup for starting up the pump.

A ¾-in. connection is provided on the pump suction downstream of the block valve. It is attached permanently to a gas pressurizing source, such as fuel gas or nitrogen. Obviously, the connection has to be made to a source that can provide gas at pressures above the bubble point of the liquid being pumped.

A ¾-in. connection is provided on the pump discharge, upstream of the block and check valves, and it is connected to the nearest flare header to vent fuel gas and/or hydrocarbon vapor during commissioning. A double block and bleed are provided on the discharge to avoid problems of freezing in cryogenic service.

The recommended procedure for starting up the pump is as follows:

1. Close the pump suction and discharge block valves.
2. Open the ¾-in. block valve labeled A in Fig. 2.
3. Open the fuel-gas/nitrogen globe valve labeled B, which will bring the system pressure up to the level that prevents flashing (this would correspond to the bubble-point pressure of the liquid). The pressure gage shown indicates the system pressure.
4. Open the suction block valve. Liquid in the suction line will not yet be able to enter the pump, due to zero pressure differential.
5. Open slightly the ¾-in. globe valve labeled C, which leads to the flare. As the gas evacuates the suction, liquid will enter the pump without flashing. It is important that Valve C be opened slowly to prevent flashing — opening it too fast or too far may cause gas to leave the system faster than liquid entering it, causing the system pressure to drop. The pressure on the gage should be kept very close to the bubble-point pressure.
6. The pump is full when a liquid/vapor cloud comes out when Valve D is cracked open slightly.
7. Once the pump is full of liquid, close Valves B, A, D and C in that order.
8. Start the pump, and open the discharge valve to the system.

Choosing plastic pumps

Plastics—especially fluoropolymers—have exceptional corrosion resistance to a wide range of chemicals, and thus are well suited for making pumps.

Edward A. Margus, *Vanton Pump & Equipment Corp.*

Typical plastic horizontal centrifugal pump used to handle corrosive chemicals   Fig. 1

☐ Industrial plastic pumps are often specified when corrosion is a problem. Pumps can be made from a wide variety of engineering plastics, and one of them can usually handle a corrosive fluid in question. Plastics are also used for erosion—many plastics offer erosion resistance superior to metals. And when zero contamination of product streams is needed, plastic is specified. Many plastics perform excellently when subjected to thermal, hydraulic and mechanical shock.

Some engineers dismiss plastic pumps because their mechanical properties are lower than those of common structural metals such as low-carbon steel. However, pump designers can engineer units that take into account the plastic's properties. These pumps are often thought to be not rugged enough, but this is not so. In fact, except for fiberglass-reinforced plastic, the plastics in pumps are unfilled, unmodified, unplasticized, virgin resins, which are high-quality to meet industrial needs.

However, plastic pumps cannot be used under all conditions. Working temperatures cannot be much above 225°F. Above such temperatures, loss of mechanical properties and stress-cracking corrosion may hinder performance. Also, there are limits on flowrate and head. This varies with each manufacturer's line, but for one line the limits are 1,000 gpm and 190 ft.

Plastic pumps used in plants are often centrifugals. Three types are used: horizontal, horizontal self-priming, and sump. Fig. 1 shows a typical horizontal type. The impeller, casing and heavy-sectioned sleeving over the shaft are made of engineering plastics.

Plastic as structural material

To better understand how plastics are used in constructing pumps, we will discuss how plastics behave under load. Not only are these materials weaker than metals, but they behave differently.

Fig. 2 shows a typical stress-strain curve for a low-carbon steel. Up to the proportional limit, stress is proportional to strain; the graph is linear. Designers keep stresses below the proportional limit. Here, the metal springs back to shape once the load is removed.

Plastics do not have an elastic region, and, thus, do not show linear behavior (see Fig. 3). Consequently, there is no fixed relationship between stress and strain, even at low loads. Also, for plastics:

■ When the load is released the material will not snap back into its original position. Permanent deflection takes place.

■ Under sustained load, the material creeps. Steel does not.

■ Temperature greatly affects behavior. For steel, its behavior will not change much below 500°F.

However, as noted, designers have found ways to cope with such behavior in constructing pumps. Much data exists on the behavior of engineering plastics. Nonetheless, owing to the nonlinear behavior of plastics, numbers cannot be plugged into formulas to accomplish a design, as with metals. Therefore, only highly qualified designers have the expertise to create workable pumps.

Materials of construction

Several types of tough, industrial plastics are used to manufacture pumps:

■ *Vinyls*—The most common are polyvinyl chloride (PVC) and chlorinated PVC (CPVC). PVC has good chemical resistance and is an excellent choice for service temperatures to 130°F. CPVC has an advantage over PVC in that it withstands temperatures to 225°F.

Service temperatures for pumps should not be confused with those listed in tables by plastics manufacturers. Maximum temperatures listed by manufacturers are those at which the plastic still retains its basic shape but may have lost its mechanical properties. Pump makers are more conservative and only list temperatures at which the plastic can carry the load imposed on it.

■ *Polypropylene*—This plastic has one of the lowest densities of the widely produced engineering plastics (its

Originally published November 12, 1984

CHOOSING PLASTIC PUMPS

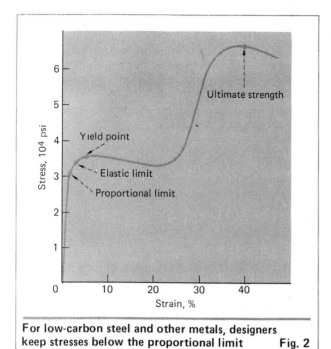

For low-carbon steel and other metals, designers keep stresses below the proportional limit — Fig. 2

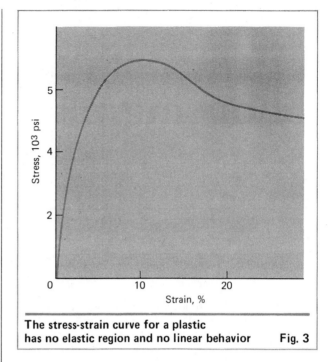

The stress-strain curve for a plastic has no elastic region and no linear behavior — Fig. 3

specific gravity is 0.90), and a relatively good stiffness. Thus, it has a good strength-to-weight ratio. Polypropylene offers broad chemical resistance to acids, caustics and solvents. It is subject to attack by strong oxidizing agents.

■ *Fluoroplastics*—These plastics have excellent corrosion resistance. Polytetrafluoroethylene (PTFE) is perhaps the most inert compound known. Polyvinylidene fluoride (PVDF) has better mechanical properties than PTFE; it is stronger, stiffer, and less subject to elongation under load. It withstands temperatures to 300°F, and has excellent chemical resistance, even to strong oxidizers.

Two modified PTFEs, created to serve for special tasks and well suited for use in pumps, are now covered.

FTFE offers some improved properties over the virgin resin

	PTFE	FTFE
Compressive creep, %		
78°F, 2,000 psi, 24 h	16	7
500°F, 600 psi, 24 h	30	10.5
Hardness, Rockwell R	15–17	27–29
Compressive stress (73°F, 1% strain)	600	1,200
Lubricated dynamic wear factor, K^*	$20,000 \times 10^{-10}$	$2\text{--}5 \times 10^{-10}$
Operating temperature, °F	–400 to +500°F	–400 to +500°F
Chemical resistance	Almost inert	Almost inert

*Special company test, where $K = R/(PVt)$

K is a relative number; the higher its value, the faster the wear rate, $(in^3)(min)/(lb)(ft)(h)$

R is radial wear, in.

P is pressure, psi

V is velocity, ft/min

t is time, h

Improved PTFE

1. A proprietary filled PTFE (FTFE)—Rulon, made by Dixon Industries Corp., Flow Control Product Unit (Bristol, R.I.)—designed as a superior material for sleeve bearings. PTFE is not a suitable bearing material unless it is fully contained to prevent cold flow. Properties of PTFE and FTFE are compared in the table.

2. Ethylene-chlorotrifluoroethylene (ECTFE) copolymer—Halar, made by Allied Corp., Plastics Div. (Morristown, N.J.)—which was designed to overcome PTFE's drawbacks in strength and fabricability. (Although PTFE is a thermoplastic, it does not melt enough to be injection-molded. It must be formed by the same methods used for sintering metals, then machined into shape.)

ECTFE was formulated to be an excellent structural material, as well as have excellent coating properties and abrasion resistance. Also, ECTFE was designed to be formed using conventional molding methods. Mechanical properties of PTFE, ECTFE and the filled fluoropolymer are compared in Fig. 4. The properties of PVDF do not appear in the figure, but they will be mentioned.

■ The ultimate tensile strength (Fig. 4a) is the stress at rupture under tension. For FTFE, the manufacturer did not provide a number since sleeve bearings are not loaded under tension. Still, the value for FTFE is lower than for the other two polymers. For PVDF, the value is about 4,500 psi.

■ The flexural (bending) modulus (Fig. 4b) measures stiffness or resistance to bending, such as a load imposed on a cantilever beam. Since no plastic has a stiffness near that of steel, the value of this modulus is critical. It is important because pump impellers and casings are subject to bending stresses (as well as to shear, tension and compression). The modulus for PVDF is the same as that of ECTFE. The modulus for ECTFE is about three times those for PTFE and FTFE.

■ Creep (Fig. 4c) is a measure of stretching under a

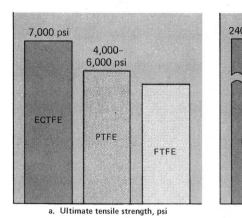
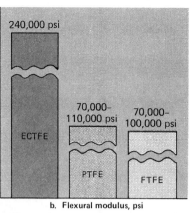
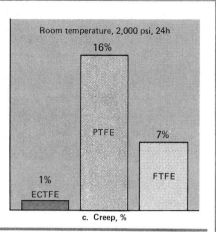

Mechanical properties of some fluoroplastics that are used in making industrial pumps Fig. 4

load as a function of time. This is a serious problem for highly loaded structural parts, and ECTFE has a low level of creep—only 1%. (For PVDF, this is about 2%.) While 1% would horrify a bridge designer, a manufacturer of plastic pumps can work around such a figure. Moderate creep is not a serious problem with sleeve bearings if they are properly contained—there is simply no place for the material to go. Pump designers try to work at maximum stress levels below 2,000 psi.

From Fig. 4, ECTFE (with PVDF close behind) looks like a superior material, and FTFE does not. However, when chemical resistance is considered, FTFE looks much better. Fig. 5 shows that ECTFE has, in general, a lower temperature rating than do PTFE and FTFE.

The temperature ranges in Fig. 5 may be high for use in pumps. For gaskets, where the plastic is completely supported and under little stress, temperatures may be realistic. But for use in structures, such as impellers and casings, one pump maker limits the working temperature to 225°F, even for the best plastics.

Further, the chemical resistance of these materials varies. As noted, virgin PTFE is almost inert. In developing FTFE, the manufacturer tried to retain this chemical resistance, and the two plastics have about the same corrosion resistance. ECTFE's chemical resistance is excellent, but it is not quite as inert as PTFE. ECTFE does, however, resist almost every chemical found in industrial service to 225°F. Only a few chemicals can attack this material; among them are dioxane at room temperature. PVDF also offers excellent resistance.

ECTFE vs. PVDF

Since both of these are top candidates for plastic pumps in severe service, a brief comparison of their properties is presented:

Chemical resistance—Both offer excellent resistance here, altough neither is as inert as pure PTFE. Particularly at elevated temperatures and high concentrations of corrodents, ECTFE has an edge over PVDF in resisting oxidizing acids and hydroxides. In general, ECTFE may have a slight advantage in terms of working temperatures. PVDF, on the other hand, resists bromine better. Both plastics show some permeability to liquid bromine, but PVDF's absorption rate is about one-tenth that of ECTFE.

Mechanical properties—PVDF is somewhat stiffer under tensile and bending loads, and has slightly higher tensile strength. However, ECTFE has better impact resistance. The coefficient of thermal expansion is about the same for the two fluoroplastics.

Formability—ECTFE may be slightly more versatile, and has the added advantage of producing an exceptionally smooth surface during injection molding. This contributes to its high resistance to abrasion and erosion.

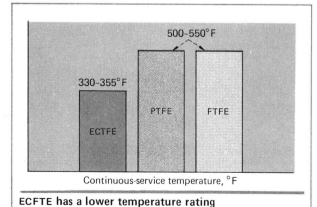

ECFTE has a lower temperature rating than do these two other fluoropolymers Fig. 5

The author

Edward A. Margus is vice-president of engineering for Vanton Pump & Equipment Corp., 201 Sweetland Ave., Hillside, NJ 07205; tel: (201) 688-4216. With the company since 1951, he and other staff members have pioneered the use of industrial-grade plastics for chemical pumps. He serves on the boards of directors of media and natural-resources groups. He has held numerous positions in the machine-tool and tooling industries. Margus holds B.S. and M.S. degrees in mechanical engineering from New Jersey Institute of Technology.

Part III
Positive-displacement pumps

Selecting positive-displacement pumps
Predicting flowrates from positive-displacement rotary pumps
Alternative to Gaede's formula for vacuum pumpdown time

SELECTING POSITIVE-DISPLACEMENT PUMPS

Here is a guide for process designers, and users of such machines. The authors describe several of the most commonly used reciprocating and rotary pumps, and tell where to apply them and how to specify and select them.

Donald J. Cody, Craig A. Vandell, and **Don Spratt,** Stearns Catalytic Corp.

Positive-displacement (PD) pumps employ fixed displacement (i.e., the same volume/revolution) and indefinite pressure to move fluids. There is little or no leakage or recirculation and these pumps are self-priming. Some are well suited for viscous, hazardous or slurry services.

Certainly, prime considerations are: When to use PD pumps?; When to choose them over centrifugals?; and, Which type is best? To answer these questions, we will look mainly at process considerations. Economics are dealt with only briefly at the end of this report, due to space limitations.

Fig. 1 shows where PD machines fit relative to each other and to centrifugals. The figure defines the maximum boundaries of PD pumps and high-speed centrifugals. It presents a minimum boundary for centrifugal pumps, so as to show where the best fits are for PD pumps. For centrifugals, the operating range is the area above the line in Fig. 1; for all others, the range is below the line. The break in the ordinate indicates gal/h, to emphasize the low-flow regimes of PD pumps. These boundaries are nominal and are an effort to illustrate the usual application ranges. They are not intended to eliminate any product or technology. For example, diaphragm pumps are available to handle flows of 500 gpm and above, but are shown to emphasize their metering capabilities and related low flows (gph vs. gpm).

Similarly, although most PD machines stop at about 10,000 ft of head, the boundary has been extended beyond that value with no particular limit. High-pressure intensifiers and hydraulic pumps, for example, are used in great quantities in some industries (but are not covered in this report). These devices are employed in hydraulic systems at pressures to 5,000 psi.

We have selected the beginning of centrifugal-pump territory as being at a specific speed of 1,000 or an optimum efficiency of about 75%. Specific speed is a dimensionless design parameter that depends upon speed, flowrate and head per impeller.

Obviously, it is sometimes desirable to use centrifugal pumps in lower flow regimes, where efficiency is not a primary consideration but the intent is rather to suggest a reasonable boundary. Centrifugals are not self-priming (except for special designs), and are unsuitable for viscous service — 1,000 Saybolt Seconds Universal (SSU) is about the end of the line. In short, the emphasis is on low-flow processes — the ones that typically confront the chemical-process design engineer.

Definition of terms

To clarify the discusssion, some terms used in this report will now be defined. Commonly known terms — such as maximum allowable speed and rated discharge pressure — are not included. Knowing the wording of a definition is often necessary for the testing of a PD pump, prior to its being accepted. A misunderstanding of a certain term can result in a test being done improperly. Thus, refer to the standards listed below when specifying a test.

Definitions are taken from standards of the American Petroleum Institute

An article on predicting flowrates for rotary positive-displacement pumps begins on p. 98.

84 POSITIVE-DISPLACEMENT PUMPS

(API). Referred to are Standard 674 (reciprocating) [1], Standard 675 (controlled-volume) [2] and Standard 676 (rotary) [3] for Positive-Displacement Pumps. The API Standards are recommended reading because of their widespread international use in the chemical process industries for design, selection and testing.

Also, Standards of the Hydraulic Institute [4] are useful for design information, terminology, testing and installation practices. The API definitions:

Acceleration head is the pressure change due to changes in velocity in the piping system. It is an important factor in the application of reciprocating pumps because of the pulsating nature of the flow in some pump suction lines, as well as net positive suction head required (NPSHR), vapor pressure, and head required to overcome suction-line losses. (For additional information on acceleration head, refer to Hydraulic Institute standards.)

Rated capacity of a rotary pump is the quantity of fluid actually delivered per unit of time at rated speed, including both the liquid and dissolved or entrained gases, under stated operating conditions. In the absence of any vapor entering or forming within the pump, the rated capacity is equal to the volume displaced per unit of time, less slip.

Rated differential pressure is the difference between rated suction pressure and rated discharge pressure.

Net positive suction head available (NPSHA) or suction lift (when the source-liquid-level is below the pump inlet) is the total suction pressure, excluding system acceleration head, available at the pump suction connection, minus the vapor pressure of the liquid at pumping temperature.

Net positive suction head required (NPSHR) is the total suction pressure, including pump internal acceleration head, required by the pump at the suction connection, minus the vapor pressure of the liquid at pumping temperature.

(NPSHR and NPSHA are normally in psi. It is the responsibility of the vendor to determine the NPSHR.)

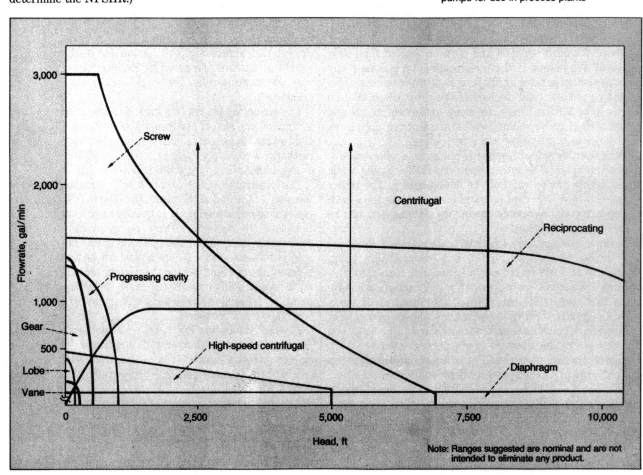

Figure 1 — This map helps in the selection of positive-displacement pumps for use in process plants

SELECTING POSITIVE-DISPLACEMENT PUMPS

Pump efficiency (also called pump mechanical efficiency) is the ratio of the pump power output (hydraulic horsepower) to the pump power input.

Volumetric efficiency is the ratio of pump suction capacity to pump displacement, expressed as a percentage.

A *controlled-volume pump* is a reciprocating pump in which precise volume control is provided by varying the effective stroke length. Such devices also are known as proportioning, chemical-injection or metering pumps. In a *packed-plunger pump*, the fluid is in direct contact with the plunger. In a *diaphragm pump*, the process fluid is isolated from the plunger by a hydraulically actuated diaphragm.

Lost motion is a means of changing displacement of a constant-stroke pump by altering the effective stroke length during each cycle. This may be done mechanically or hydraulically.

Turndown ratio is the rated capacity divided by the minimum capacity that can be obtained while maintaining specified repetitive accuracy and linearity.

Linearity describes change of flowrate in response to change of capacity adjustment. It is defined as the difference between actual flowrate divided by capacity adjustment, related to a straight line, and is expressed as a percent of rated capacity.

Flow repeatability, expressed as a percent of rated capacity, describes the reproducibility of pump flowrate under a given set of conditions when capacity setting is varied and then returned to the setpoint being tested.

Steady-state accuracy is the flow variation expressed as a percentage of mean delivered flow under fixed-system conditions. Steady-state accuracy applies over the turndown ratio.

Rotary pump is a PD pump consisting of a casing containing gears, screws, lobes, cams, vanes, plungers or similar elements actuated by relative rotation between the driveshaft and the casing. There are no separate inlet and outlet valves. These pumps are characterized by their close running-clearances.

Displacement of a rotary pump is the volume displaced per revolution of the rotor(s). In pumps incorporating two or more rotors operating at different speeds, the displacement is the volume displaced per revolution of the driving rotor. Displacement depends only on the physical dimensions of the pumping elements.

Rated speed of a rotary pump is the number of revolutions per minute of the rotor(s) required to meet the specified operating conditions. In pumps incorporating two or more rotating elements operating at different speeds, the rated speed is the speed of the driving rotor.

Slip is the quantity of fluid that leaks through the internal clearances of a rotary pump per unit of time. Slip depends on the internal clearances, the differential pressure, the characteristics of the fluid handled, and, in some cases, the speed.

RECIPROCATING PUMPS

Piston and plunger

Perhaps the most common and certainly one of the oldest types of PD pumps is the reciprocating type. This pump creates suction (it actually fills its cylinder) on one stroke of a piston or plunger, and discharges the pressurized fluid on the reverse stroke. There are many different classes and subclasses of reciprocating pumps, as shown in Fig. 2; the two major ones are power and direct-acting.

Both are offered with pistons, plungers or diaphragms. Power pumps can be either single- or double-acting while direct-acting pumps can only be double-acting. The power pump can have one, two or more pistons while the direct-acting pump can have only one or two pistons; both can be either horizontal or vertical.

A further description of each of these classes follows:

A *power pump* (see Fig. 3) is a reciprocating pump consisting of a power end and a liquid end, connected by a "frame or distance" piece. The power end contains a mechanism that transmits energy from a rotating shaft (on a motor, engine or turbine) to pistons or plungers by means of a crankshaft, crossheads and connecting rods. The liquid end consists of cylinders, pistons or plungers, and valves.

A *direct-acting pump* (see Fig. 4) is a reciprocating pump that consists of a piston-powered drive end connected directly to a liquid end. Power is transmitted to the liquid end by action of a motive fluid at the drive end. A direct-acting pump can use steam, air or unburned fuel gas as the motive fluid. Subclasses are:

Horizontal or *vertical* — referring to the orientation of the piston rods.

Single-acting pumps (see Fig. 3) are those reciprocating devices that alternately take suction and discharge on one side of a piston or plunger. This results in one suction and one discharge stroke per cycle or revolution of the crankshaft.

Double-acting pumps (see Figs. 4 and 5) alternately take suction on one side of the piston and discharge pressurized fluid on the opposite side. This results in two suction and two discharge strokes per cycle or revolution of the crankshaft. Direct-acting pumps are always double-acting.

Piston pumps (see Fig. 5) have a large cylinder (the piston) mounted at the end of the piston rod. The O.D. of the piston has sealing or piston rings that limit leakage from one side of the piston to the other. These pumps usually require a replaceable liner within the cylinder, since the cylinder is subject to wear. Piston pumps should not be used when the differential pressure across the piston exceeds 2,000 psi. Such pressure differentials can cause deflection within the pump, with resultant leaks across the rings, excessive ring wear, and other problems.

Plunger pumps (see Fig. 3) have a smooth rod (the plunger) attached to the plunger rod. A plunger is smaller in diameter than a piston. It is sealed from leaking to the outside by stationary packing in the stuffing box. Plunger pumps are used for high discharge pressures (i.e., for differential pressures above 2,000 psi) and can attain heads in excess of 10,000 ft.

Diaphragm pumps will be described later on.

Simplex pumps have one cylinder and, thus, one discharge and one suction stroke per cycle or crankshaft revolution for a single-acting pump, and two of each type of stroke for a double-acting pump.

Duplex pumps have two cylinders, each with a piston driven off the main crankshaft. This essentially doubles the number of suction and discharge strokes over the simplex pump.

Multiplex pumps have larger multiples of cylinders. The advantages to using pumps with multiple cylinders will be discussed later on.

When to apply piston/plunger pumps

Before considering the specifics of the pump design, one must look at the type of system that requires a piston or power pump. Fig. 1 shows that the most obvious application is for low flows and high heads. Reciprocating pumps can operate at discharge pressures to 10,000 psi.

A primary advantage of a reciprocating pump is that high pressures can be achieved at low piston velocities. This makes reciprocating pumps particularly suitable for handling high-viscosity fluids and, with major modifications, abrasive slurries (such as coal slurries).

Another use is for metering or proportioning. Such metering pumps are modifications of power pumps. Capacity is measured in gallons per hour and the stroke is adjustable from 0–100% of capacity, either manually or automatically, usually while the pump is running. These metering pumps use ball check-valves, similar to those found in diaphragm pumps.

The power pump's high overall efficiency (85–94%) often makes it competitive with centrifugal and rotary pumps for services where hydraulic capabilities overlap. This high efficiency favors specifying power pumps, especially due to the cost of energy.

Direct-acting pumps have many of the same advantages of power pumps, plus a few specific ones that make them particularly suited to several applications: They have an infinite range of capacities and discharge pressures that can be varied independently. Capacity is proportional to speed, and the speed is controlled by throttling the motive fluid — steam, air or fuel gas. Thus, direct-acting pumps are well suited whenever a wide variation of flow is required.

Direct-acting pumps can accomplish two different tasks at once. The obvious one is pumping. The other is that the drive end of the pump acts as a throttle valve. These pumps do not depend on steam or gas expansion to accomplish pumping, but act only as reducing valves. Therefore, the heat consumed is almost entirely due to radiation loss and can be minimized using insulation. When high-pressure steam or fuel gas must be throttled, a direct-acting pump should be considered. This results in free pumping, since the motive fluid would normally be throttled through a valve.

Direct-acting and power pumps are self-priming, making them advantageous for suction-lift applications. A slow-speed reciprocating pump in cold-water service at sea level should be able to operate with a total suction lift of up to 22 ft.

This must be derated for higher elevations (due to lower atmospheric pressures), higher fluid temperatures (due to higher vapor pressures), and fluids having a vapor pressure higher than water's (see Fig. 6).

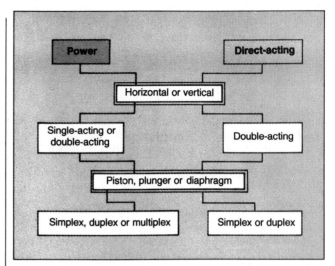

Figure 2 — Various types of reciprocating pumps

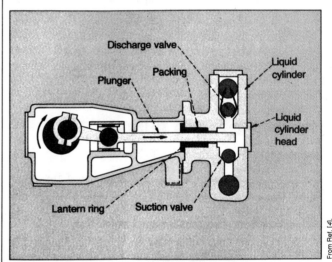

Figure 3 — Power pumps are driven by rotating shafts that are connected to motors, engines or turbines at the power end

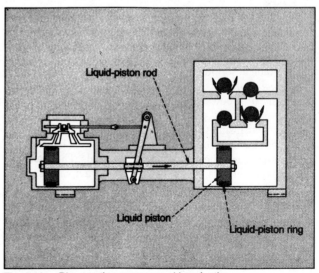

Figure 4 — Direct-acting pumps are driven by the action of a motive fluid—steam, air or unburned fuel gas

SELECTING POSITIVE-DISPLACEMENT PUMPS

The theoretical maximum suction lift for cold-water service is 34 ft (14.7 psia converted to feet of head). The difference between this and actual lift represents losses in the system, such as those due to pipe friction, valves, and velocity head. Design of an individual pump could also derate suction lift; the manufacturer should be consulted in all suction lift applications.

Where *not* to use piston/plunger pumps

While it is obvious that reciprocating pumps should not be applied outside of the hydraulic range shown in Fig. 1, there are other limitations on their use.

The primary one is that they cannot be used wherever a completely pulseless flow is required. As will be discussed later, reciprocating pumps, by their very nature, can create some fairly severe pulsations in the piping system. These pulsations can be greatly reduced with pulsation dampeners, but can never be completely eliminated.

When reciprocating pumps are used in viscous service, there is a corresponding decrease in maximum allowable speed for them to operate satisfactorily. A pump operating at 100% speed at 250 SSU would have to have its speed reduced by 35% if viscosity were increased to 5,000 SSU. As noted before, a decrease in speed results in a decrease in volumetric flow; thus, flows are reduced. This decrease limits the flow range in Fig. 1 even further.

Similarly, pumps at higher fluid-temperatures must sometimes be reduced in speed for proper operation. For both viscous and high-temperature service, the pump manufacturer should be consulted as to proper application.

Finally, reciprocating pumps should not be used when they are uneconomical in relation to either rotary or centrifugal units. Economics will be briefly mentioned later on.

Specification, evaluation and selection

A major part of this is understanding how the pump will be used in the pumping system. A pump is not just a piece of machinery that raises the pressure of the fluid; it is part of an overall system. To provide adequate service life and minimize maintenance, the system must be well understood.

Fig. 7 [5] shows a pumping system with recommended components and fluid velocities. This is a conservative design that yields satisfactory pump performance and long life, and minimizes maintenance. Let us discuss major components of the system:

The suction vessel is equipped with a weir plate and vortex breaker to minimize turbulence and air bubbles that can enter the suction piping. The feed and bypass lines enter the suction vessel below the minimum liquid level, to eliminate gas bubbles — entrained air or gas in the fluid increases compressibility and can severely affect the performance of a reciprocating pump.

The suction line is sized to limit the fluid velocity to 1–3 ft/s. This low velocity reduces the acceleration head of the system, maximizing the NPSHA to the pump. The long-radius elbow minimizes pressure drop and maximizes the NPSHA.

The eccentric reducer decreases the line size from the larger suction-pipe diameter down to the inlet size of the pump. The orientation of this reducer is always such that an air pocket cannot form in the pipe. In fact, the entire piping system should be arranged so that no air pockets can form. If the piping must be designed with a high point where air pockets can form, then this point should be vented. As mentioned earlier, air in the system can restrict performance or, even worse, keep the pump from working at all.

The suction stabilizer usually contains internals such as bladders, diaphragms, choke tubes or baffles. It is used in front of the pump to reduce the pressure pulsation in the pipe and acceleration head in the system. These pulses can result in damage to the pump, pipe and supporting structures.

The figure shows a discharge pulsation dampener mounted on a leg. A dampener is required here to reduce the extremely high-pressure pulses that can occur in the discharge piping of the system. The one in the figure uses a bladder that requires an air or nitrogen charge for operation. The bladder expands or contracts with pulsations in the line, equalizing pressure variations.

A flowthrough type of dampener, mounted inline, usually consisting of a sphere without internals, is also available. There are advantages and disadvantages to each type of dampener. The pump manufacturer should be consulted in selecting this device.

The relief valve protects the pump in case there is a restriction, upset, or closed valve in the discharge system. Being a PD device, the reciprocating pump will attempt to bring the fluid to the discharge pressure. As this pressure increases, the pump will continue to operate against it, until the motor overloads, something breaks or, preferably, the relief valve opens.

Relief valves should be set to fully relieve at a pressure no greater than the pump cylinders' maximum allowable working pressure, but at no less than 110% of rated discharge pressure. The relief valve must be able to handle the pump's rated capacity when the valve is fully open and at a pressure no higher than 10% above its set pressure.

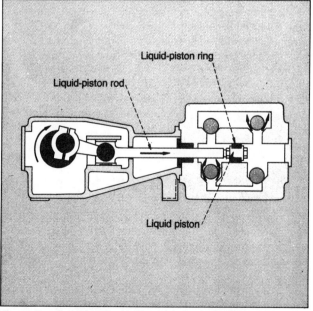

Figure 5 — Compare this double-acting piston power-pump with Figure 4, a double-acting steam pump. Both are horizontal

88 POSITIVE-DISPLACEMENT PUMPS

A direct-acting pump requires special consideration in relief-valve sizing in that this type of pump has a stall pressure defined as:

Stall pressure = (gas piston dia.²/liquid piston dia.) × maximum inlet gas pressure + maximum liquid suction pressure (1)

and a ram pressure defined as:

Ram pressure = (gas piston dia.²/liquid rod dia.) × maximum inlet gas pressure (2)

If either of these values exceeds the maximum allowable working pressure of the pump or that of any discharge-system component, then a properly sized relief valve must be provided. If a relief valve is required as a result of ram pressure, then the system must be designed so that the discharge of the relief valve vents to outside the pump block valves.

The startup and capacity-control bypass is constructed so that the pump can be started under no load and then be gradually loaded. Reciprocating pumps can be started without this bypass, but use of it is easier on the pump and driver. The bypass also allows for capacity control when required for the process. The line velocity for this section should be 5–15 ft/s so there is no excessive backpressure.

The discharge line should be equipped with a check valve and a discharge gate-valve. Line velocity in the discharge should be limited to 3–10 ft/s or no more than three times the suction-line velocity.

As has been mentioned several times, limiting fluid velocity in the lines, especially in the suction, is critical to the satisfactory operation of the system. See Fig. 8, showing the maximum allowable suction-line velocity vs. crankshaft speed for duplex, triplex and quintuplex pumps.

Acceleration head

The major reason for limiting suction-line fluid velocity is to maximize NPSHA for the pump. NPSHA is calculated in the same way as for rotary or centrifugal pumps with one major exception — a term must be included for acceleration head:

$$NPSHA = H_a - H_{vp} \pm H_{st} - H_{fs} - H_{ac} \quad (3)$$

where:

$NPSHA$ = net positive suction head available, ft

H_a = absolute pressure acting on fluid in suction vessel or sump, ft

H_{vp} = vapor pressure of fluid being pumped at pumping temperature, ft

H_{st} = static height above or below centerline of cylinder; positive when the pump is taking suction from an elevated vessel, and negative in a suction-lift situation, ft

H_{fs} = piping friction loss, ft

H_{ac} = acceleration head, ft

Due to the alternating suction and discharge strokes of a reciprocating pump-cylinder, fluid is constantly being accelerated and decelerated. The resultant velocity fluctuations become pressure pulsations within the piping. As Eq. (3) indicates, a change of pressure on the suction side changes NPSHA for the system. The below-average pressure in the

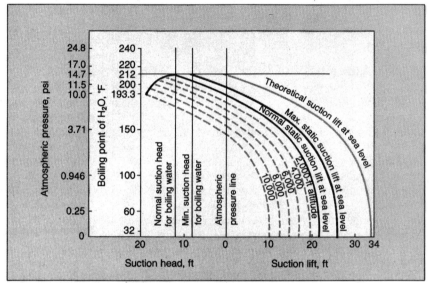

Figure 6 — Theoretical water suction-lift corrected for altitude, atmospheric pressure and fluid temperature

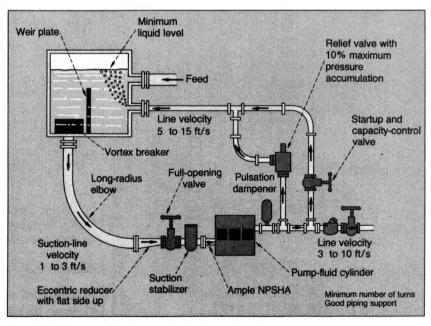

Figure 7 — Recommended system design for reciprocating pumps

suction line is the acceleration head; it can be determined from the following empirical formulation:

$$H_{ac} = LVNC/32.2K \qquad (4)$$

where:
H_{ac} = acceleration head (pressure reduction below average pressure), ft
L = actual pipe length (not equivalent length), ft
V = average fluid velocity in pipe (should be limited to 1–3 ft/s), ft/s
[V = gal/min × 0.3205/area (in²)]
N = pump crankshaft speed, rpm
C = constant dependent upon type of pump:
0.200 for single-acting duplex
0.115 for double-acting duplex
0.066 for single- or double-acting triplex
0.040 for single- or double-acting quintuplex
0.028 for single- or double-acting septuplex
K = factor representing the reciprocal of the fraction of theoretical acceleration head that must be provided to avoid a noticeable disturbance in suction line:
1.4 for relatively incompressible fluids (e.g., deaerated hot water)
1.5 for water, amines, glycols
2.0 for most hydrocarbons
2.5 for relatively compressible fluids (e.g., hot oil, ethane)

The value of H_{ac} must be kept as low as possible to maximize NPSHA. Eq. (4) shows that as the average fluid velocity increases, H_{ac} also increases. This is why low fluid velocities are required in the suction piping. Average velocity also increases as the pump speed increases; therefore, H_{ac} varies with the square of the pump speed. As a result, one must determine whether limiting the pump speed is advantageous when H_{ac} is high.

Since H_{ac} varies directly with the length of the suction piping, it is obvious that this length must be kept to a minimum. H_{ac} also varies with the type of pump (i.e., the constant C). In general, overall pulsation is less for pumps with more cylinders, and H_{ac}, thus, is less also. In addition, the more incompressible the fluid, the higher the H_{ac}.

The suction stabilizer has an effect on H_{ac}, since it lowers the value of L to about 10 pipe diameters [*4*], thus reducing H_{ac} to a minimum.

Pressure pulses

The discharge pulsation dampener is required to reduce the high-pressure pulses in the discharge piping system.

Fig. 9 [*5*] shows average velocity for various reciprocating pumps vs. crankshaft angle. The velocity and flowrate vary significantly from the average. If the abscissa were changed from crankshaft angle to time, the slope of the curve would become the acceleration. This acceleration multiplied by the mass and divided by the flow area of the pipe is the pressure pulse. In general, the larger the number of cylinders on a pump, the smaller the slope (or acceleration) and thus the smaller the pulse.

This can be seen in Fig. 9. The slope for the duplex single-acting pump is much greater than that for the quintuplex single-acting one. Since use of pulsation dampeners greatly reduces damage to the pumping system, items such as pump valves will have a longer life.

Other things to be considered within the pumping system are fluid properties, such as specific gravity, viscosity, amount of entrained gases, and temperature. Two examples: In general, higher specific gravities result in higher brake horsepowers; higher temperatures and viscosities will lower the maximum piston speed allowed. Any significant amount of entrained gases should be discussed with the manufacturer.

Mechanical considerations

To design the pumping system for maximum life and minimum maintenance, the piston or plunger speed of the pump must be limited. Tables I and II [*1*] give limitations for pump

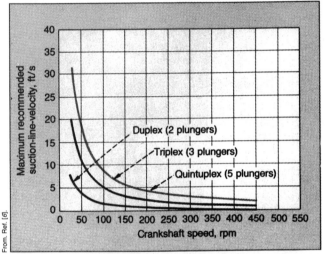

Figure 8 — Maximum allowable suction-line velocity varies with the crankshaft speed and the number of plungers

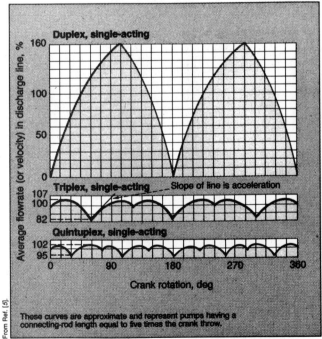

Figure 9 — For reciprocating pumps, a greater number of cylinders minimizes pressure pulse transmitted to the system

speed in rpm and ft/min for pumps of various sizes. In addition, Table I lists viscosity correction-factors that further limit the speed. Other items that could also result in limited speed are temperature, and the amount and size of solids in a slurry.

The packing and packing configuration must be considered as well. Packings are the highest-maintenance items in the pump. Packing material should be compatible with the pumped fluid and its temperature. The complexity of the packing arrangement depends on pressure, viscosity, toxicity of the fluid, and so on. Ref. [5] shows common stuffing-box designs for reciprocating pumps.

Probably the next most maintenance-intensive item in a power pump is the plunger. Its surface wears due to high-speed movement within the packing. For this reason, plungers are protected with a hard coating, such as chrome, ceramics or nickel- or cobalt-based alloys. The coating should be chosen in consultation with the pump manufacturer.

Recommended materials of construction for other parts of the pump are listed in Tables B1, A2, B3, in Appendix B of API 674. In general, I-1 and I-2 classes are for water or nonflammable services, and S-1 and S-2 classes are for flammable and corrosive services at the liquid end. For the gas-end parts of a direct-acting pump, I-1 and I-2 classes are for steam or air, while S-1 and S-2 are for flammable gases such as fuel gas.

As regards drivers for power pumps, typical ones are electric motors, steam turbines, and diesel or gas engines. The driver must be sized to meet the maximum specified operating conditions, including gear and coupling losses. Any anticipated variations in process conditions, such as in temperature, pressure or fluid properties, that may affect driver sizing should be accounted for. Starting conditions (loaded or, preferably, unloaded) also affect the driver.

For motor-driven units, the driver nameplate-horsepower should be at least 110% above the maximum anticipated horsepower (usually at the relief-valve setpoint), figuring in gear and coupling losses. Power-pump speeds range from 20 to 450 rpm, yet motor speeds are more typically 1,200 to 1,800 rpm. This means, of course, that some type of speed reducer must be used.

API 674 limits the use of gear motors to 25 hp and below. Typically, V-belt drives are limited to 100 hp, while gears are used for higher-horsepower units. For variable-flow motor-driven units, variable-speed motors are commonly used at 60 hp and below, while hydraulic couplings and eddy-current clutches are applied at higher horsepowers.

Integral (one-piece) multi-V-belts should be used when belt drives are needed. However, caution should be exercised in specifying belts, since they are troublesome.

Belts should be nonsparking (or antistatic) if the electrical area-classification requires an explosionproof or totally enclosed fan-cooled (TEFC) motor. If more than one integral multi-V-belt is required, the manufacturer should furnish belts of matched length. The service factor [a factor that accounts for rapid changes in the load on the belt] should be at least 1.5 for multiplex plunger pumps; 1.6 for duplex double-acting piston pumps; and 1.75 for duplex single-acting pumps. V-belt speeds should not exceed 5,000 ft/min. To aid in maintenance, the manufacturer should supply a positive-belt-tensioning device that has an adjustable base with guide and hold-down bolts, a tensioning screw and lock bolts. Motors in V-belt service should be designed for the side-load imposed by the V-belt drive on the motor (i.e., the belt is 90 deg. to the shaft).

All gears should conform, at a minimum, with Amer. Gear Manufacturers Assn. (Arlington, Va.) AGMA 420. If the rated horsepower of the gear is above 1,000 hp, consider specifying one that conforms to API Standard 613.

Steam turbines, if used, should be designed to a well-recognized specification, such as API Standard 611. The turbine should be sized to continuously provide the horsepower to drive the pump at any of the operating conditions at the required speed and steam conditions.

Table I — Maximum allowable speed ratings for power pumps in continuous service as given by API Standard 674 [1]

Stroke length, in.	Speed rating			
	Single-acting plunger-type pumps		Double-acting piston-type pumps	
	rpm	ft/min	rpm	ft/min
2	450	150	140	46.5
3	400	200	—	—
4	350	233	116	77.0
5	310	258	—	—
6	270	270	100	100
7	240	280	—	—
8	210	280	—	—
10	—	—	83	138
12	—	—	78	156
14	—	—	74	173
16	—	—	70	186

Note: For an intermediate stroke length, the maximum speed shall be interpolated from the numbers in the table.

As a guide for viscosities above 300 Saybolt Seconds Universal (65 mm²/s) at pumping temperature, speeds should not exceed the following percentages of the speeds given in Tables I and II.

Viscosity, SSU	Percent of speed
300–1,000	90
1,000–2,000	80
2,000–4,000	70
4,000–6,500	60
6,500–8,000	55

Table II — Maxium allowable speed ratings for direct-acting pumps in continuous service, again given by API Standard 674

Stroke length, in.	Speed rating	
	cycles/min	ft/min
4	52	35
6	44	44
8	38	51
10	34	56
12	30	61
14	28	65
16	26	69
18	24	72
20	22	75
24	20	80

SELECTING POSITIVE-DISPLACEMENT PUMPS

If the service has a high suction pressure, be sure that the manufacturer has strengthened its entire pump to handle the additional loads created by the high suction pressure. Not only might the suction flange and manifold have to have their design pressures increased, but the crankshaft and bearings, crosshead pin, and other parts might have to be redesigned.

This is a consideration for single-acting pumps where the suction pressure exceeds 5% of the maximum discharge pressure. On double-acting pumps, where the suction pressure exceeds the 5% value, a tail rod (a rod the same diameter as the piston rod) should be added to the piston or plunger opposite the piston rod. This will balance the areas and, thus, the forces

Inspection and testing

Once the process and mechanical design have been set, next decide which inspections and tests are required to ensure that the pump train is of desired quality and will perform properly. Indeed, most reciprocating pumps are purchased to the manufacturer's standard test and inspection procedures.

For critical or marginally designed systems (such as those with a minimal NPSHA), inspection and test procedures must be stringent and agreed on by buyer and seller before issuance of a purchase order. What if the pump is to be placed in a critical service in which the user stands to lose a great deal of money if downtime or delayed startup occurs? Stringent inspection and testing in the manufacturer's shop should uncover most problems, so that such expenses are minimized.

Inspection covers records kept — such as material certification, purchase specifications for all subvendor-supplied

Table III — API Standard 674 acceptance criteria for reciprocating pumps

Characteristic	Tolerance, %	
	Power pump	Direct-acting pump
Rated capacity	+3, −0	+3, −0
Rated power (at rated pressure and capacity)	+4	—
NPSHR	+0	+0

data, quality-assurance and running-test data, surface and subsurface inspections, and final dimensional inspection. API Standard 674 offers guidelines on the types of inspection required, and on acceptance criteria.

Once the inspection areas have been chosen, one must figure out whether any of these inspections should be "witnessed" (this implies a "hold point"), "observed" (the engineer or inspector is notified about three days prior to inspection, but the inspection is not held, awaiting the inspector), or just "recorded". This decision usually depends on the criticality of service, cost of the equipment, remoteness of installation, and other variables that would warrant spending more money for the increased observation of the manufacturing process.

The running tests, such as performance and NPSH tests, also should be agreed on before a purchase order is issued. The desired test procedure should be specified, including such items as how long the test is to be run, how many data points are required and, perhaps most importantly, what are the acceptance criteria. A full proposed test procedure should be requested from the manufacturer with its quote.

The proposed procedure should be evaluated against the specification and the desired acceptance criteria to see whether the test will furnish the needed information. If not, one should negotiate with the manufacturer to change the test. It cannot be overemphasized that this negotiation should be done before the purchase order is placed. The test procedure, especially acceptance criteria, should be made a part of the purchase order.

Recommended tests can be found in the Standards of the Hydraulic Institute or API Standard 674. As a minimum, the performance test for power pumps should be run long enough to obtain complete information on speed, suction and discharge pressures, capacity and power. For variable-speed drives, the manufacturer should test over the entire speed range. For direct-acting pumps, variables should be proven at speeds of 25, 50, 75, 100 and 125% of rated speed. The NPSH test should show that, at rated speed with the NPSHA equal to the quoted NPSHR, pump capability is within 3% of the noncavitating capacity. Table III contains acceptance criteria taken from API Standard 674 for different pumps and tests.

For economic reasons, NPSH tests should usually be performed only when the NPSHR is within 2 ft of the NPSHA. Also, nonwitnessed performance tests with required review of results prior to shipment are adequate for most services. Witnessed tests should be done only on critical services, expensive pumps, and pumps to be used under unusual process conditions, such as high suction pressures.

Evaluation

All critical items—such as performance conditions, metallurgy, horsepower, type of packing, and packing arrangement—should be evaluated. If a vendor has not quoted to the specification while others have, then that vendor should either be dropped or brought up to specification so the evaluation will be equal. The final choice is ultimately the manufacturer who is technically acceptable (it is seldom that a manufacturer complies completely with the specification) at the lowest possible cost (including an economic evaluation).

Diaphragm pumps

The diaphragm pump is a reciprocating pump in which the "pumpage" is isolated and moved by a flexible diaphragm or diaphragms. Fig. 10 shows a double-acting pump with ball valves at the suction and discharge ports. Only one of the two pumping chambers has been exposed.

The amount of fluid displaced by diaphragm pumps is somewhat limited (see Fig. 1); therefore, capacity varies from about 0.1 gal/h to 100 gpm and beyond, using multiple diaphragms. This is similar to multiplexing in a reciprocating pump. Pressures range from low (about 100 psi) for mechanically driven elastomeric diaphragms, to high (about 50,000 psi) for hydraulically driven metal diaphragms. Flow direction is established by inlet and discharge check valves. Other features such as built-in relief valves and diaphragm-failure detectors may be included.

The inherent good volumetric efficiency and zero slip make the diaphragm pump an excellent controlled-volume (metering) device, with the ability to precisely set and control flow.

It is available in a wide variety of materials: synthetic elastomers such as neoprene, butyl or nitrile rubber; fluoroelastomers such as polytetrafluoroethylene (PTFE); or metals (usually corrosion-resistant types). All diaphragms have good fatigue properties. Check valves and seats are used, and are made of hard or corrosion-resistant materials (including synthetic gems).

When to apply diaphragm pumps

There are several situations where these devices are ideal:

They are, of course, excellent for controlled-volume pumping. Manual or feedback control can be used. Steady-state accuracy is good (±1% or better over the entire turndown range). So are linearity (±3% or better over the turndown range) and turndown ratio (10:1 or better, and this can be done while the pump is running). Repeatability is good, too (±3% over the turndown range). Since there is no slip, the flow is insensitive to pressure changes. Diaphragms are self-priming.

Diaphragms can be used to isolate the pumpage, making them good for hazardous, toxic, pure and similar services. Double diaphragms have a buffer fluid that further enhances isolation. The pump can be isolated, too. Since the diaphragm can be driven hydraulically, valves can be remotely located (as can the entire diaphragm assembly) to permit isolation of the suction and discharge. This is usually done for safety where the pumpage is lethal, explosive, hot, radioactive, and the like.

These pumps also are used for slurries. Since the valves can be separated from the diaphragm by a hydraulic line, buffer fluids can be employed to keep slurry material away from the diaphragm (using a surge leg), and restrict the slurry to the valve area.

Where *not* to apply diaphragm pumps

When there are pressure and capacity restraints — these units are not suitable for high flowrates (see Fig. 1). Generally, they are unsuitable for pulseless flow. If this is desirable, a rotary pump should be considered. However, pulsation dampeners can resolve the problem, unless subtle amounts of pulsation are intolerable.

There is a maximum permissible viscosity of about 3,500 SSU. Valving may be the limiting factor for viscosity. PD pumps usually handle the viscous liquids quite well if they can get past the valves. And although these pumps can accept considerable amounts of entrained gases, if the pumpage or hydraulic fluid has too much entrained gas, performance (particularly as a metering pump) can be seriously affected.

Specification, evaluation and selection

NPSHA or suction lift can be calculated using any number of suitable texts. These quantities should be evaluated before vendors are asked to quote, to be sure the vendors are not being asked to go beyond the state of the art.

Like all self-priming pumps, diaphragm pumps should be able to operate at least 5 psia of NPSHA or 22 ft of H_2O (19 in. of Hg) of suction lift. However, acceleration head must be accounted for. This head should be calculated prior to asking for quotes. The design should be adjusted accordingly if customary restraints are exceeded. Many times, acceleration-head problems can be solved with charged accumulators, which contain bladders filled with gas.

Since velocity head is part of NPSHR, too high a head can rob the machine of static pressure, leading to noise, erratic performance and possible damage. Velocity head is dictated by the process designer or the pump designer. Therefore, evaluation of inlet piping and inlet valves, both in the pipe and in the pump, is needed.

Of course, pumping requirements must be considered. These include service conditions (flows, pressures, normal and upset conditions, etc.), fluid properties, and construction features. In addition to the valves and diaphragms discussed above, the bidder should adequately describe the housings, bearings, seals, gaskets, plungers, controls and auxiliaries.

The preciseness of flow control and its type (manual, hydraulic, etc.) should be conveyed to the bidder. The bidder should fully describe the flow-control mechanism. Metallurgical selection must be reviewed. Tables 1 and 2 from API Standard 675 are excellent sources for material combinations.

Further, repair of new parts by the manufacturer should be in accordance with good practice such as set forth in ASTM and other standards and codes. Drivers need to be fully described, including area classification and nameplate

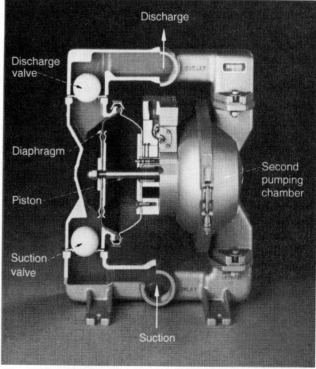

Figure 10 — Double-acting diaphragm pump with one half of it cut away to show diaphragm, and inlet and discharge valves

power-rating to accommodate relief-valve accumulation.

Testing, whether done to the manufacturer's or the purchaser's standard, should be specified. Witnessing or observing, documenting and reporting should be mutually understood. Hydrostatic and performance tests are the most common ones to be witnessed.

ROTARY PUMPS

There are five main types: screw, progressing-cavity (which is actually a type of screw pump), gear, lobe, and vane. These PD pumps are distinguished from reciprocating types in that the displacement of the pumped fluid is caused by a rotary motion of the rotating element(s) of the pump, rather than a reciprocating motion.

Rotary pumps differ from centrifugals in that the amount of fluid displaced by each revolution is largely independent of speed. A small amount of slip occurs, which depends on discharge pressure, viscosity and pump speed.

All rotary pumps consist of one or more rotating members or rotors that move within a stationary casing or stator. As fluid enters through the inlet port, it is trapped in a volume created between the rotor and the stator. Good pump design should allow this volume to increase in size uniformly as the rotor element (gear teeth, vanes, lobes, etc.) passes the inlet port. This will ensure good suction performance of the pump, i.e., a low NPSHR.

As the rotor continues its motion, there should be a point where the trapped volume is closed to both the inlet and outlet ports. If this is not so, the pump is not a truly PD device and must rely on fluid inertia. In a well-designed pump, the trapped volume also should remain constant until the fluid reaches the outlet port. There, the volume should uniformly decrease in size to expel the fluid.

Screw pumps

As the name suggests, screw pumps are rotary pumps that use one, two or three screws to advance fluid from inlet to discharge. Since, in many cases, rotary pumps can be operated much faster than reciprocating ones, they offer a sizable flow range, nominally up to 3,000 gpm, as well as high discharge pressures to 3,000 psi. They do not require valves, which makes them suitable for rigorous-NPSH, viscous, and non-Newtonian service due to their good inlet geometry.

The most common of the single-screw pumps is the progressing cavity (see Fig. 11). It has a single-helix rotor running in a double-helix stator of twice the pitch. Usually the stator is a nonmetallic material that provides a seal along the helix. This restrains slip and permits entry of some solid particles. Slip is not eliminated and will take place during pressure and viscosity upsets, thus providing some protection to this device.

Such pumps are ideal for high-viscosity and non-Newtonian fluids. However, heads and flowrates are a good deal more limited than those of two- and three-screw models. The longer the pump, the longer the seal along the helix and the higher the pressure-rise capability.

Two-screw machines typically employ timing gears to prevent intimate contact and thus permit the pumpage to establish a lubricating film between the rotors. However, such construction leads to higher slip. Opposed helices are used to balance thrust. Three-screw pumps (Fig. 12) use two helices for bearing support, and are untimed.

Slip varies inversely with viscosity in screw pumps. Thus, inlet sizing is set at maximum viscosity, and displacement (due to lowest volumetric efficiency) at minimum viscosity.

Where to apply screw pumps

Screw pumps are used when a self-priming device is needed. They readily handle liquids of good lubricity (e.g., oil at 20 cP) and high viscosity. The maximum viscosity is about 4 million SSU.

Single-screw models handle viscous, non-Newtonian, corrosive pumpage. They are good where slip is useful during upset conditions. The elastic-stator seal permits higher recirculation at higher backpressures. Metal-to-metal screw pumps make for a stiffer system. Some abrasives can be pumped through them. They should be used if a change in the properties of the pumpage can occur, i.e., change in viscosity of a non-Newtonian liquid. They provide pulseless flow at low speeds.

Double-screw pumps with timing gears are suited for nonviscous, poorly lubricating, abrasive and corrosive liquids. They provide pulseless flow at high or low speeds.

Three-screw devices accept viscous, non-Newtonian liquids, and provide pulseless flow at high or low speeds.

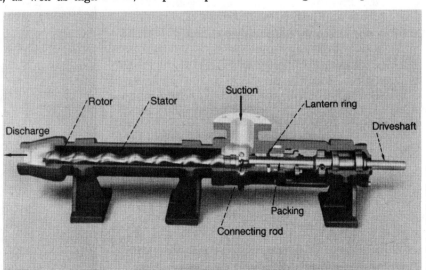

Figure 11 — Progressing-cavity is a single-screw pump, which is ideal for pumping non-Newtonian and viscous materials

Where not to apply screw pumps

Do not use these pumps for severe slurry service in which particles are hard or too numerous. In some designs, low lubricity of the pumpage limits use. If the viscosity of the pumpage remains below 10 cSt for any extended period, lubricity should be pursued in detail with the manufacturer.

Specification, evaluation and selection

See the remarks on diaphragm pumps. Also: Since, for non-Newtonian fluids, viscosity can vary nonuniformly with shear rate and time, problems can arise.

Instruments used to measure viscosity of a non-Newtonian fluid may not necessarily correlate with what is actually going on inside a pump. Although time usually is not a factor, it can be when material is recycled. For these pumps, shear rate can vary at different places, as well as over the pumping cycle. Also, non-Newtonian fluid properties can be undesirably altered (damaged) if care is not exercised.

Another consideration in non-Newtonian regimes is that often the flow is laminar, not turbulent, usually causing higher-than-expected pressure drops. Sometimes the situation may seem paradoxical, such as when lower viscosities result from higher shear rates. This leads to the conclusion that increasing the velocity could reduce energy losses. However, since getting the material into the pump is usually the problem, and flow is usually laminar, this drop in viscosity with shear is often of little use. Attention must be given to inlet piping as well as to pump inlet design. Numerous articles are available on viscous flow through pipes, valves and fittings. However, in non-Newtonian flow, relating apparent viscosities often requires direct testing and help from the manufacturer.

If the pumpage is a slurry, the settling velocity needs to be evaluated. A description of the entire pumping loop is most valuable to the bidder. Viscosity variations due to temperature can be handled nicely if the range is understood at the outset.

For single-screw pumps, the rotor moves in an orbital path around the drive-shaft centerline, requiring the use of a sliding, double universal joint for proper transmission of power. The U-joint designs should be weighed for durability and maintenance, particularly in slurry service or where commingling of pumpage and joint lubrication can destroy lubricant properties.

For two-screw machines, the merits of timing gears, including external vs. internal, should be weighed. If the pumpage is not clean or lacks sufficient lubricity, externally located and lubricated timing gears are preferable.

As for all PD machines, flow can be controlled by altering speed. The effects of speed, however, also have to be weighed against viscosity, especially for apparent viscosity for non-Newtonian fluids.

For selection of metals and alloys, see Table B-1 from API Standard 676, Positive Displacement Pumps — Rotary.

Multispeed drivers (at constant torque) may be desirable when viscosity varies as part of the normal and/or upset process conditions.

As for inspection and testing, the extent of witnessing, observing, documenting and reporting should be clear.

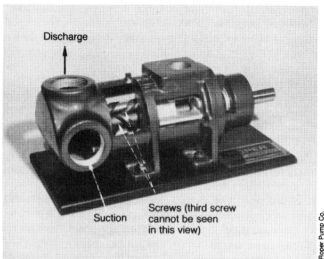

Figure 12 — Three-screw pump employs two helices for bearing support

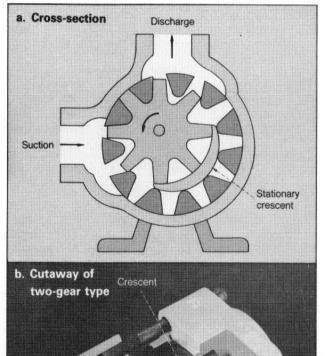

Figure 13 — Internal gear pump is used to handle oils, water-base fluids, glycols, phosphate esters, emulsions

SELECTING POSITIVE-DISPLACEMENT PUMPS

Gear, lobe and vane pumps

Gear, lobe and vane pumps are used where relatively low flows (600 gpm or less) and low differential pressures (400 psi and below) are required.

They are especially suited for viscous and non-Newtonian fluids, where the loss of efficiency makes centrifugals uneconomical.

Gear pumps — In this type of rotary pump, the volume between the rotor and stator is created by the gear teeth. Fig. 13 shows an internal type; Fig. 14, an external type. There is necessarily a small clearance between the gear teeth and the casing, but sliding contact exists between the gears. For this reason, external timing gears may be used when pumping non-lubricating fluids.

Lobe pumps — In this type of rotary pump (see Fig. 15), the trapped volume is created by the lobes of the rotor. Unlike the teeth of the gear pump, these lobes are incapable of driving one another, and external timing gears are mandatory. There should be no contact between the individual rotors or between the rotors and the stator if the pump is to be used on nonlubricating fluids.

Vane pumps — These are generally classified into two groups — rigid rotor (see Fig. 16) and flexible rotor. The rigid rotor relies on centrifugal force acting on the vanes to maintain the seal between the rotor (i.e., the vanes) and stator. This type of rotary pump self-compensates for wear on the vanes, which slide in slots in the rotor. There is a close clearance between the endplates of the housing and the rotor.

Variable displacement may be obtained by altering the eccentricity of the rotor within the casing, either with a manual control lever or by various other means — electrically, pneumatically or hydraulically.

The flexible vane depends upon the elastomeric properties of the rotor material. This pump is also self-compensating for wear on the vanes, but care must be exercised in specifying this type of pump in fluids that may have a detrimental effect (e.g., hardening or swelling) on the elastomeric material.

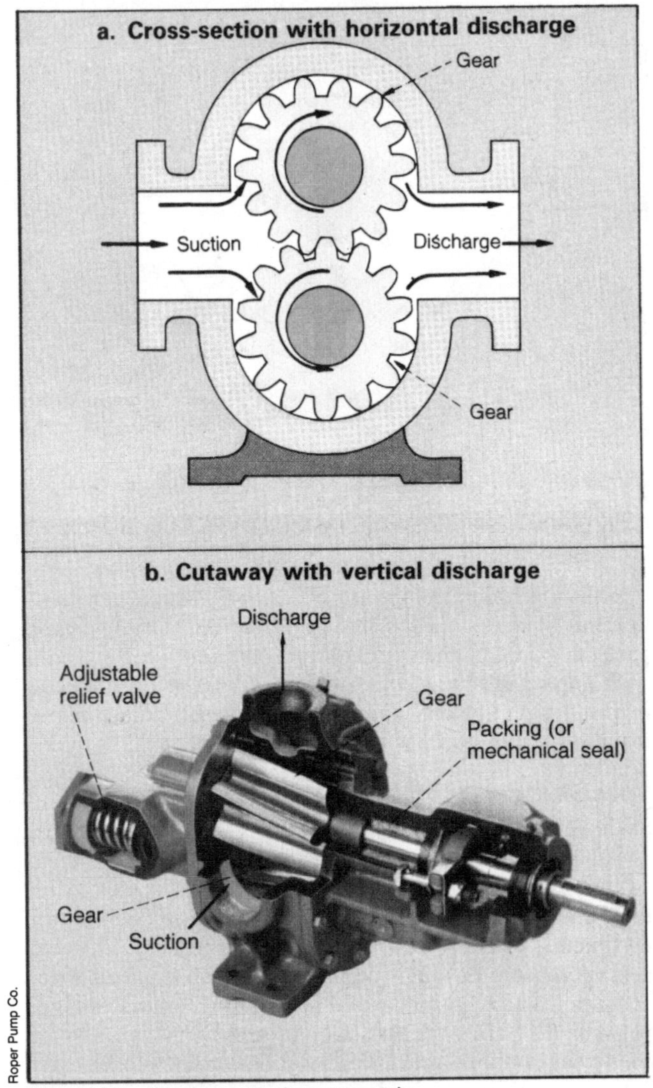

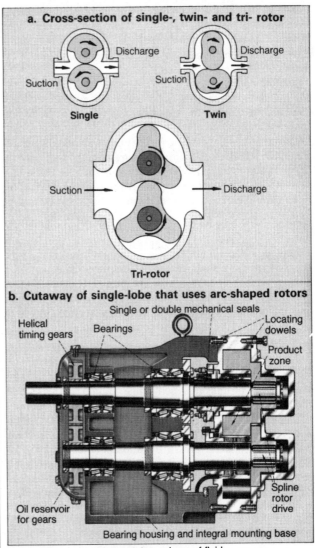

Figure 14 — External helical gear-pump is used in applications similar to those of internal gear

Figure 15 — Lobe pump entraps volume of fluid between lobe and casing. Rotors should not contact each other

When to apply rotary pumps

Rotary pumps are especially suited to pumping viscous, lubricating fluids and are almost universally applied to lube- and seal-oil systems when flowrates are in these pumps' range of operation. Nonlubricating fluids (e.g., water, fuels, various chemicals, etc.) can be handled by lobe or screw pumps, provided that external timing gears are used to eliminate sliding contact between the rotors.

Volumetric efficiencies of rotary pumps can be as high as 95–98%, with overall efficiencies of 70–95%, thus making them preferable to centrifugals when the cost of electricity is high. Over the low-flow ranges typical of rotary pumps, centrifugals generally have efficiencies of 30–60%.

For example, compare the costs of electricity per year for both types of pumps. Assume that power is 5¢/kWh for an 8,000-h operating year. For a 300-gpm, 200-psi, 40%-efficient centrifugal pump operating on water, the cost is $26,100/yr. A rotary pump operating under the same conditions would have an efficiency of, say, 85%. The cost of electricity would be $12,300/yr, representing a saving of $13,800/yr.

Rotary pumps are of relatively simple construction and, provided their hydraulic end is contained in a cartridge type of design, they are easy to maintain. In a cartridge setup, the pumping element comes out in one piece, complete with seals, etc. This makes servicing easy.

Suction and discharge flows are relatively free of pulsation, a continual problem with reciprocating pumps. This eliminates the need for expensive suction and discharge pulsation-dampeners.

Since they are PD devices, rotary pumps are ideally suited to non-Newtonian fluids, although care must be exercised in sizing the pumps and drivers for such service. Since the viscosity of non-Newtonian fluids varies with agitation or shear rate, the expected viscosity under the actual running conditions must be found. This is apparent viscosity, and is meaningful only if stated at a given shear rate. As noted before, empirical data or actual test results are usually the only way of correctly sizing a pump and its driver for non-Newtonian fluids.

Rotary pumps are generally self-priming, unlike centrifugals (unless these are specifically designed for the purpose), and can therefore be used where a high concentration of gases is present or where a suction lift is required. For nonlubricating fluids where the pump must be self-priming, a liquid film must be present between the externally-timed rotors. This effects a seal and allows the priming to take place. It may be difficult to self-prime a pump that has stood idle for long, thus allowing liquid to drain from the rotors.

The theoretical maximum height of the pump centerline above the liquid level (the suction lift) depends on several factors: atmospheric pressure (altitude); fluid vapor-pressure and, therefore, temperature; and suction losses (pipe friction, valve losses, and velocity head). The pump manufacturer should always be consulted if a suction-lift application is encountered.

When *not* to apply rotary pumps

The obvious restraint is, again, where flow and pressure are outside of the hydraulic coverage chart, Fig. 1. However, this should not deter the user from seeking any special designs that may be available — for example, for high pressures.

Rotary pumps are not suited for handling slurries or any fluid containing even small amounts of solids, especially if these particles are abrasive. Due to the close running-clearances in screw, gear, lobe and vane pumps, they should not be used when solids are in the pumped fluid, unless there is a suitable strainer in the suction piping. The manufacturer should always be consulted in such cases.

Also due to these close clearances, there is somewhat of a lower temperature limitation on these pumps than on centrifugals. The slip factor should always be taken into account when proposing to pump very-low-viscosity fluids with rotary pumps. Liquid gases can be pumped with vane pumps, but other types of rotary pumps are generally unsuitable.

Rotary pumps should not be used if any flow variation is required, unless a variable-speed drive is employed. The exception to this is with variable-eccentricity vane pumps, where flow can be varied, as mentioned earlier.

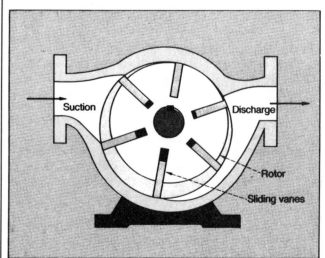

Figure 16 — Vane pumps can be used to handle liquid gases — other rotary pumps cannot maintain a seal between the stator and rotors

Although rotary pumps are of relatively simple construction, and maintenance is limited to replacing the wearing parts, high maintenance costs are invariably encountered with appreciable wear. Sometimes, all of the internal and external parts that can wear must be replaced. In such cases, installing a new pump is often more economical.

Specification, evaluation and selection

Some vendors make many different types of rotary pumps, and these firms should be initially consulted for recommendations on the right type. The NPSH available at the centerline of the suction nozzle should be stated in the inquiry documents. Review of this figure is necessary to avoid asking vendors to go beyond the state of the art. Also, vendors should be made aware of any required margin between NPSHA and NPSHR. As already noted, service conditions and fluid properties must be considered. Be careful about solids, entrained gases and non-Newtonian fluids. State any special requirements, such as suction lift.

SELECTING POSITIVE-DISPLACEMENT PUMPS

The purchaser should describe in the specification or inquiry documents any special materials of construction, bearing-housing details, bearings and bearing life, seals, gaskets, etc. that are dictated by the fluid properties or the environment in which the pump will be used.

Note whether the unit allows for easy maintenance of the wearing parts. Consideration should be given to pumps with cartridge-type internals that are simpler to maintain.

As mentioned earlier, the inlet should be designed so that the trapped volume increases uniformly in size until the inlet port is closed. Also, the inlet port should at no time be open to the outlet port. A constant volume should be maintained between the closing of the inlet port and the opening of the outlet port. The trapped volume should then decrease uniformly after the outlet port is opened. Proposals should be checked to see that the above are closely adhered to. Caution: Some cheaper designs do not always follow these rules.

Any unusual site conditions should be noted in the inquiry, along with whether the pump is to be installed indoors or outdoors, etc. When bids are evaluated, bidders' metallurgical selection should be reviewed for conformance with specifications, suitability for liquid pumped, etc. See Appendix B, Table B-1, from API Standard 676 for material combinations.

The type of driver required should be stated (electric motor, steam turbine, etc.). A margin of 110% of the shaft horsepower at the pressure corresponding to the relief-valve setting is normally considered to be the minimum required motor-nameplate rating.

ECONOMIC EVALUATION

Once pumps have been selected to meet technical needs, an economic analysis needs to be done to find the least-costly pump equipment. With the ever-increasing cost of energy, the efficiency and time of operation (operating cost) can be significant. So is the technical ability to evaluate manufacturers' claims of efficiency.

Thus, capital costs alone are becoming less meaningful. Often, maintenance costs are elusive, and probably are best determined from one's own company's experiences. Even then, unless well documented, experience can become distorted; for example, someone may be unwilling to concede that equipment was misapplied. Depending upon the desired depth of the evaluation, consider some of the following:

1. Capital cost (include related equipment such as utilities).
2. Operating cost (assign an energy cost and be sure efficiency claims are verifiable).
3. Maintenance cost.
4. Schedule of payments (find net present worth).
5. Cost of money (assign inflation and interest rates).
6. Tax impact.
7. Operating time.
8. Downtime (include criticality of service, and installed or warehoused spares).
9. Lease vs. purchase.
10. National purchase agreements.
11. Original-equipment-manufacturer (OEM) discounts.
12. Quantity discounts.
13. Cost of spare parts (either at time of purchase of original equipment or later).
14. Delivery of software and hardware related to all schedules and the cost impact.
15. Payback period.
16. Service labor and costs (precommissioning, commissioning, repair, availability, location, stocking of spares, etc.)
17. Warranty protection.
18. Escalation.
19. Liquidated damages.

Due to space limitations, these items cannot be covered here. However, there are numerous publications on engineering economics, as well as the knowledge of your firm's cost engineers, that can be helpful.

References

1. Amer. Petroleum Institute, API Standard 674, Positive Displacement Pumps — Reciprocating, 1st ed., API, Washington, D.C., May 1980.
2. Amer. Petroleum Institute, API Standard 675, Positive Displacement Pumps — Controlled-Volume, 1st ed., API, Washington, D.C., Mar. 1980.
3. Amer. Petroleum Institute, API Standard 676, Positive Displacement Pumps — Rotary, 1st ed., API, Washington, D.C., Sept. 1980.
4. The Hydraulic Institute Standards, 14th ed., Hydraulic Institute, Cleveland.
5. Henshaw, Terry, Reciprocating Pumps, *Chem. Eng.*, Vol. 88, No. 19, p. 105, Sept. 21, 1981.
6. Henshaw, Terry, Keys to Optimum Power-Pump Operation, *Plant Eng.*, Dec. 6, 1976, p. 173.

The authors

Donald J. Cody is staff machinery engineer for Stearns Catalytic Corp., P.O. Box 5888, Denver, CO 80217. Tel: (303)692-2099. He has been with Stearns for 17 years. Mr. Cody has spent considerable time in precommissioning, startup and troubleshooting of machinery in the field, and in manufacturers' plants for inspection, etc. He is a member of API task forces for Standards 611, 612, 671, and 673, for which he is chairman. He holds a B.S.E.E. degree from the University of Colorado and is a P.E. in Colorado and Alaska.

Craig A. Vandell is machinery engineering supervisor at the above address. Tel: (303) 692-3516. His recent work has included providing mechanical support to the proposal and presentation for a North Slope exploration and production facility. He is responsible for equipment sizing optimization, pricing, scheduling and manpower. He has been with the firm since 1976. Previously, he was rotating-machinery engineer for Procon, Inc. Mr. Vandell holds a B.S.M.E. degree from the University of Illinois, and has completed an arctic engineering course as a step toward getting his P.E. license in Alaska.

Donald Spratt is machinery engineer at the above address. Tel: (303) 692-2870. He has nine years' experience in refinery and chemical-plant construction. He is responsible for large rotating machinery. His duties include establishing budgets and schedules, writing specifications, and generating mechanical flow-diagrams. He has three years' service in U.K. Government Research. He obtained a polytechnic diploma in mechanical engineering from Lanchester Polytechnic, Rugby, Warwickshire, U.K., and is a graduate member of the Institute of Mechanical Engineers.

PREDICTING FLOWRATES FROM POSITIVE-DISPLACEMENT ROTARY PUMPS

Michael L. Dillon, Kenneth A. St. Clair, and **Philip H. Kline,**
Robbins & Myers, Inc.

Performance depends on so many variables that each pump must be calibrated for specific operating conditions. How these variables affect performance is explained, and a method for programming the calibrations is presented.

Flowrates from positive-displacement rotary pumps are uniform and predictable at particular speeds, if other variables are held constant. These variables are pump speed, differential pressure, viscosity, temperature, materials of construction, and the dimensional fit of the pumping elements. Flowrate is proportional to speed at zero differential pressure if viscosity and pump speed are low enough to prevent cavitation.*

Just as higher viscosity increases friction loss in piping, it also tends to reduce slip ("blowby") in positive-displacement pumps. The apparent viscosity of non-Newtonian fluids varies with shear rate, which in pumps is related to pump speed. Therefore, the performance (including cavitation and slippage effects) of a particular pump handling a specific fluid can be unique for each pump operating speed, even though other pumps of the same model and type are supposedly identical. This uniqueness is a function of fittings between mating pump parts. Minute differences in the manufacturing of pump parts make the slip index of each pump different.

Although it is possible to calculate theoretical performance for fixed capacity, speed, viscosity and differential pressure, the most accurate method of predicting pump performance is to rely on actual flow tests of a fluid pumped at two representative speeds, and to generate equations that describe the actual performance. The equations will include the effects of viscosity and fitting tolerances on performance. By interpolating between these two base-performance equations, one can predict the flowrate for a specific fluid at any particular pressure and speed.

The method presented here will enable engineers to predict flows for applications in which flowmeters would be unreliable or too costly to install.

Two of the factors affecting flowrate from positive-displacement rotary pumps — viscosity and differential pressure — warrant extended discussion of the role they play in making it difficult to predict pump performance.

Viscosity's impact on flowrate

Until the rotor, piston, lobe, gear or vane closes behind the fluid and applies pressure to it, the pump can only create a void. The quantity of fluid that flows into the void will depend (as with any orifice) on the fluid's viscosity, the differential pressure across the opening, and the entrance loss (K factor), which reduces (due to turbulence, friction, vena contracta, etc.) the flow theoretically possible.

If the fluid's vapor pressure and the flooded-head friction loss at the pump's suction are negligible, the maximum differential pressure the pump can create by opening a void will be approximately 14.7 psi at sea level. As long as the pressure drop between the suction port and the entrance to the pumping element does not exceed 14.7 psi, the fluid will fill the void, and full-displacement flow will result. If, however, this pressure drop is higher, part of the void

For a guide to selecting positive-displacement pumps — both rotary and reciprocating — see the preceding article.

*All rotary positive-displacement pumps operate by trapping a quantity of fluid between a stationary element and one or more moving elements — gears, helical rotors, screws, lobes or vanes. As the rotating element moves within the stationary element, or in conjunction with another rotating element, the pump delivers a set quantity of fluid per rotation. Discounting differential pressure, when a pump operates at speeds at which the available net positive suction head (NPSH) exceeds the pressure drop across the pump suction, a boost in pump speed will produce a proportional increase in the pumping rate of thin fluids.

Originally published July 22, 1985

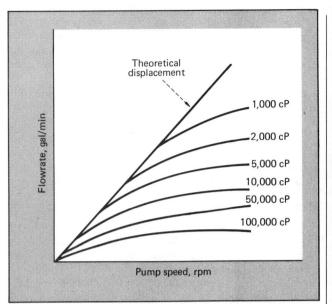

Figure 1 — Typical curves indicate deviation from theoretical displacement for Newtonian fluids of various viscosities at 1 atm NPSH

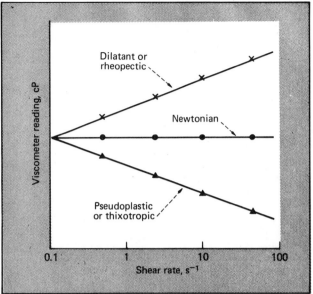

Figure 3 — Viscosities of thixotropic and rheopectic fluids change with amount and time-length of shearing

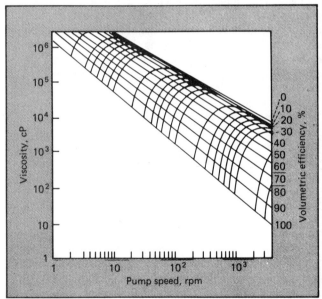

Figure 2 — Typical plot yields maximum pump speed for handling Newtonian fluids of various viscosities at a given available NPSH

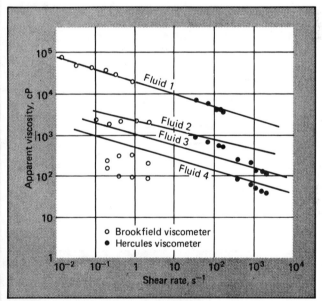

Figure 4 — Typical changes in apparent viscosities of non-Newtonian fluids with varying shear rate

will be filled with vapor, which will be compressed and condensed upon the application of pressure. This will cause the flowrate to deviate from the theoretical-capacity-vs.-pump-speed curve. As Fig. 1 indicates, the more viscous the fluid, the higher the pressure drop and the larger the flowrate deviation.

For a specific pump model handling Newtonian fluids of various viscosities, curves can be developed to indicate the maximum speed at which the pump can be operated for a particular net-positive suction head (NPSH). Fig. 2, which shows such a curve, indicates the effect of both speed and viscosity (of Newtonian fluids only) on theoretical capacity.

The viscosity of a non-Newtonian fluid will increase or decrease as it is sheared in the pump and piping. The relationship between shear rate, which is rarely linear, and viscosity is usually an inverse one. With increasing shear rate, the viscosity of a pseudoplastic fluid decreases and that of a dilatant fluid increases.

The viscosity of a thixotropic fluid decreases and that of a rheopectic fluid increases with the intensity of shearing and the length of time during which shearing takes place (Fig. 3). Even though velocity is the same at two different points in a pipe, the viscosity of such a fluid will be higher (or lower) at the downstream point because it has been subjected to flow shearing stresses for a longer interval. This dependence of shearing on time makes specifying piping difficult, and invalidates laboratory data obtained by recirculating such a fluid.

100 POSITIVE-DISPLACEMENT PUMPS

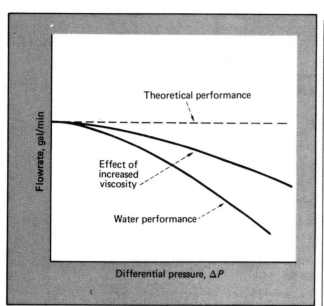

Figure 5 — Slip increases with viscosity and differential pressure, reducing flowrate from positive-displacement rotary pump

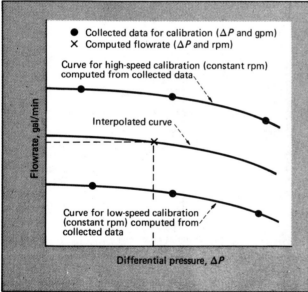

Figure 6 — Graphical representation explains calibration of program and calculation of flowrate from pumps

Average shear rates inside positive-displacement pumps are a function of speed and design constants. From the calculation of shear rate, the viscosity in a pump can be estimated. From a plot of shear rate and viscosity (as in Fig. 4), an apparent viscosity can be selected for determining pump horsepower and volumetric efficiency for specific operating conditions.

Differential pressure's effect on flowrate

Some slip ("blowby") occurs in all positive-displacement rotary pumps. Higher pressure at the discharge than that at the suction will force fluid back through the sealing lines created between a pump's rotating and stationary elements.

The amount of slip depends not only on the pressure differential but also on the number of seal lines and the viscosity of the fluid. (As mentioned, higher viscosity reduces slip and enables a pump to handle higher pressures.)

Fig. 5 shows the effect of higher viscosity on pump output. However, such is the case only when the pump is operating in a viscosity and speed range that allows it to function at 100% volumetric efficiency. On the one hand, higher viscosity increases pump capacity because it reduces slip; on the other, higher viscosity reduces capacity because it increases entrance losses and decreases volumetric efficiency. Because the wear on the rotating and stationary elements increases clearances, flowrates are prone to change with time.

Gathering the data

Because of the differing effects of all the many variables on flowrate, predicting flowrates from laboratory tests is virtually impossible. Although manufacturers can estimate performance suitably enough for pump selection, none can predict flowrate accurately without data obtained under actual operating conditions. Furthermore, because pumps change with time, a quick method is also desirable for recalibrating them.

Viscosity's nonlinear effect on theoretical capacity can also be determined by an appropriate choice of pump speeds. If a pump must operate at less than 100% volumetric efficiency, the data points for calibrating the program must also be taken at the lower efficiency. If a pump is to operate at about 100%, data must be collected at about this efficiency. If it is to run at both efficiencies, data should be taken at one representative speed for each condition.

The apparent viscosity of the fluid should be approximated by means of a curve such as shown in Fig. 4, and the volumetric efficiency for the pump speed checked by a plot such as Fig. 2. The differential pressure should be regulated by a control valve in the discharge pipe.

Tachometer readings should be taken to ensure that pressure readings are taken at the same pump rotational speed. For each speed, pressure data should be collected at one point below, at one point approximately equal to, and at one point substantially beyond the estimated actual differential pressure. Flowrates can be determined using a bucket and stopwatch or a calibrated meter.

The calibration and computational procedures are graphically portrayed in Fig. 6, which illustrates how the program works when interpolating between the "calibration curves."

Program predicts flowrate

The flow diagram for a computer or calculator program that mathematically describes a pump performance curve (capacity vs. differential pressure) at two different speeds is presented in Fig. 7. After the data for all six collection points, and the differential pressure and pump speed, have been entered, the program will calculate the flowrate, and will interpolate between, or extrapolate above and below, the two resulting curves.

The test data are entered in Steps 1 and 2. Although three data pairs for each pump rotational speed can be sufficient, additional data pairs obviously will enhance the accuracy and reliability of predictions.

A second-order equation is fitted through each set of speed

1. **Enter data for first representative speed**

For RPM_1, enter: $\Delta P_{1,1}$, $\Delta P_{1,2}$, $\Delta P_{1,3}$, etc.

and: $Q_{1,1}$, $Q_{1,2}$, $Q_{1,3}$, etc.

Note: Choose RPM_1 near highest normal operating speed

2. **Compute coefficients for equation to fit first data set**

$$A_{0,1} + A_{1,1} \times \Delta P + A_{2,1} \times \Delta P^2 = Q$$

For finding coefficients via least-squares approximation, the simultaneous equations in matrix form are:

$$\begin{bmatrix} N_1 & \Sigma(\Delta P) & \Sigma(\Delta P^2) \\ \Sigma(\Delta P) & \Sigma(\Delta P^2) & \Sigma(\Delta P^3) \\ \Sigma(\Delta P^2) & \Sigma(\Delta P^3) & \Sigma(\Delta P^4) \end{bmatrix} \times \begin{bmatrix} A_{0,1} \\ A_{1,1} \\ A_{2,1} \end{bmatrix} = \begin{bmatrix} \Sigma(Q) \\ \Sigma(Q \times \Delta P) \\ \Sigma(Q \times \Delta P^2) \end{bmatrix}$$

Here, N_1 = number of data pairs (ΔP and Q) entered in first data set

3. **Enter data for second representative speed**

For RPM_2, enter: $\Delta P_{2,1}$, $\Delta P_{2,2}$, $\Delta P_{2,3}$, etc.

and: $Q_{2,1}$, $Q_{2,2}$, $Q_{2,3}$, etc.

Note: Choose RPM_2 near lowest normal operating speed

4. **Compute coefficients for equation to fit second data set**

$$A_{0,2} + A_{1,2} \times \Delta P + A_{2,2} \times \Delta P^2 = Q$$

For finding coefficients via least-squares approximation, the simultaneous equations in matrix form are:

$$\begin{bmatrix} N_2 & \Sigma(\Delta P) & \Sigma(\Delta P^2) \\ \Sigma(\Delta P) & \Sigma(\Delta P^2) & \Sigma(\Delta P^3) \\ \Sigma(\Delta P^2) & \Sigma(\Delta P^3) & \Sigma(\Delta P^4) \end{bmatrix} \times \begin{bmatrix} A_{0,2} \\ A_{1,2} \\ A_{2,2} \end{bmatrix} = \begin{bmatrix} \Sigma(Q) \\ \Sigma(Q \times \Delta P) \\ \Sigma(Q \times \Delta P^2) \end{bmatrix}$$

Here, N_2 = number of data pairs (ΔP and Q) entered in second data set

5. **Compute flowrate for RPM and ΔP of interest**

$$Q = [A_{0,1} - S_R(A_{0,2})] + \{[A_{1,1} - S_R(A_{1,1} - A_{1,2})] \times \Delta P\} + \{[A_{2,1} - S_R(A_{2,1} - A_{2,2})] \times \Delta P^2\}$$

Here, $S_R = (RPM_1 - RPM)/(RPM_1 - RPM_2)$

Figure 7 — Flow diagram outlines program for predicting flowrate from positive-displacement rotary pumps

data by means of the least-squares method in Steps 2 and 4. Such an equation can be fitted to represent almost all pump performances within the range of practical interest. If a pump's performance differs significantly from the norm, an equation of a more appropriate form must be substituted into the algorithm. In most cases, however, the equation provided will be sufficiently accurate over a reasonable range of practical interest. The three equation shown in matrix form are solved simultaneously, by any convenient means, to generate the coefficients needed to describe each curve.

The first four steps essentially "calibrate" the program. If several fluids of differing viscosities will be pumped, a set of "calibration factors" for each fluid should be stored for retrieval to compute flowrate.

In the fifth step, a linear interpolation scheme computes a new flowrate for a particular pump speed and differential pressure. It can be repeated as often as necessary. As long as the fluid's properties (viscosity, temperature, etc.) remain the same, and pump wear does not become excessive, the equation of this step can continue to be used. If, however, there is a significant change in fluid properties or pump performance, the first four steps must be repeated so as to "recalibrate" the program.

The more closely the selected "calibration" data points represent actual operations, the greater the accuracy of the predicted flowrate. Two representative pump speeds are used to allow for performance variations due to varying shear rate and volumetric efficiency.

Flowrate also depends on driver

With positive-displacement rotary pumps, flowrate consistency is very much a function of the driver's speed-holding capability. Mechanical variable-speed drives of the variable-pitch sheave type are claimed by manufacturers to have a speed-holding capability of ±1%. Direct-current motors with a silicon-controlled-rectifier (SCR) controller have a speed-holding capability of ±5%. This can be reduced to ±1–2% with the addition of an isolation transformer and tachometer feedback. Driver speed variation must be considered in gauging the program's accuracy.

The effect of turndown ratio on the accuracy of the program is more pronounced for differential pressure than for pump speed. For this reason, it is important to estimate the narrowest pressure band in which the pump will actually operate. Because the matrix equation in the program fits a parabolic curve, and because of the wide range of pressure-differential data collected for the calibration, the program may overshoot actual capacities in the mid-pressure range.

The authors

Michael L. Dillon is vice-president and general manager of Robbins & Myers Canada, Ltd. (P.O. Box 280, Brantford, Ont. N3T 5N6; tel. 519-752-5447). Previously, he served as energy market manager and Moyno oilfield equipment manager at the company's Fluids Handling Div. at Springfield, Ohio. He holds an M.B.A. in financial administration from Wright State University in Dayton, Ohio.

Kenneth A. St. Clair is an engineer in hydromechanical accessory design with the Aircraft Engine Group of General Electric Co. (P.O. Box 156301, Cincinnati, Ohio; tel. 513-243-5696). At the time this article was written, he was a project engineer in research and development with Robbins & Myers' Moyno Industrial Products Group. He holds a B.S. degree in mechanical engineering from the University of Cincinnati.

Philip H. Kline is energy market manager in Robbins & Myers' Moyno Industrial Products Group (1895 W. Jefferson St., Springfield, OH 45501; tel. 513-327-3013), responsible for planning and directing business strategies in energy-related industries. He holds a B.S. degree in mechanical engineering from the New Jersey Institute of Technology, in Newark.

Alternative to Gaede's formula for vacuum pumpdown time

Serban Constantinescu*

When selecting a vacuum pump to evacuate a vessel, one of the first considerations is the pumpdown time. The traditional way to resolve this question is to use Gaede's formula which gives the rate of pressure change in a vacuum chamber of volume V evacuated by a pump:

$$\frac{dP}{dt} = \frac{S}{V}(P - P_a) \quad (1)$$

(Note that P_a is the lowest pressure a pump can produce at which the speed drops to zero. It is generally obtained from a speed-pressure curve supplied by the pump manufacturer. S is a measure of the volume of gas removed in a unit time from a vessel of volume V. S depends on the pressure and is generally obtained from the speed-pressure curve.)

If S is constant, the equation has an easy solution:

$$t = \frac{V}{S} \ln\left(\frac{P_o - P_a}{P_n - P_a}\right) \quad (2)$$

And the pumpdown time can be easily found. However, pump speed, S, is not constant; it slows down as the vacuum increases. To account for this, two methods are used:

1. Replacing S by an average value, $(S_o + S_n)/2$, which is a constant. Eq. (2) then becomes:

$$t = \frac{V}{\frac{S_o + S_n}{2}} \ln\left(\frac{P_o - P_a}{P_n - P_a}\right) \quad (3)$$

2. Applying Eq. (2) to small pressure ranges. The time is

Nomenclature

M	Mass, lb_m/mole	
m_o	Mass of gas in cylinder, lb_m	
P	Pressure, torr	
P_a	Ultimate pressure, torr	
P_o	Starting pressure in the vessel, torr	
P_n	Final pressure in the vessel, torr	
R	Gas constant, 1,545 lb-ft/(mole)(°R)	
R_{PS}	Pump speed, rotations (or strokes)/s	
S	Pump speed, liters/s	
S_o	Pump speed at P_o, L/s	
S_n	Pump speed at P_n, L/s	
T	Temperature, °R	
t	Pumpdown time, s	
V	Volume of vessel, L	
V_o	Volume of pump chamber, L	

*Uniroyal Chemical, P.O. Box 460, Painesville, OH 44077

broken into i increments, and Eq. (2) is applied to the increments. Pumpdown time becomes:

$$t = t_1 + t_2 + \cdots + t_i \quad (4)$$

However, this method is tedious and the result is only an approximation, unless $i \to \infty$.

Note that to use Gaede's formula, a curve of pump speed and a value of the ultimate pressure are needed.

Alternative method

A better method is proposed here:

$$t = \frac{\ln\left(\frac{P_n}{P_o}\right)}{(R_{PS}) \ln\left(\frac{V}{V + V_o}\right)} \quad (5)$$

Eq. (5) is exact and is the result of applying the Boyle-Mariotte law to each stroke of the pump piston that isolates, compresses and discharges a small amount of gas to the atmosphere. (Every mechanical vacuum pump is some kind of positive-displacement device, either a reciprocating or rotary one.)

To use Eq. (5), the rotations (or strokes)/s and the pump-chamber volume are needed.

To determine t, one needs to know how many strokes a piston pump will take to lower the pressure in a vessel from P_o to a desired P_n. Assume a constant temperature.

Before starting the pump, the condition of the gas in the vessel is given by the ideal gas law.

When the pump is connected to the vessel, the equation becomes:

$$P_1(V + V_o) = \frac{m_o}{M} RT \quad (6)$$

1st stroke $\quad P_o V = P_1(V + V_o) \quad (7)$

n stroke $\quad P_{n-1} V = P_n(V + V_o) \quad (8)$

$$P_1 = \frac{P_o V}{V + V_o}, \; P_2 = \frac{P_o V^2}{(V + V_o)^2}, \; P_3 = \frac{P_o V^3}{(V + V_o)^3} \quad \text{and}$$

$$P_n = \frac{P_o V^n}{(V + V_o)^n} \quad (9)$$

$$n = \frac{\ln\left(\frac{P_n}{P_o}\right)}{\ln\left(\frac{V}{V + V_o}\right)} \quad (10)$$

Originally published May 11, 1987

Since $n = (R_{PS})t$, $(R_{PS})t$ can be substituted for n in Eq. (10) to yield Eq. (5).

Example—A volume of 30 L (air) is to be evacuated from atmospheric pressure to 0.01 torr, using a vacuum pump. From the speed-pressure curve of the pump; pump speed at 0.01 torr is $S_n = 1.2$ L/s; and pump speed at atmospheric pressure is $S_o = 2.4$ P_a is 0.006 torr. From manufacturer's data the R_{PS} is 12, and V_o is 0.2 L. Results by Eqs. (3) and (5) are compared:

$$t = \frac{30}{\frac{2.4 + 1.2}{2}} \ln\left(\frac{760 - 0.006}{0.01 - 0.006}\right) = 203 \text{ s} \tag{3}$$

$$t = \frac{\ln\left(\frac{0.01}{760}\right)}{12 \ln\left(\frac{30}{30 + 0.2}\right)} = 141 \text{ s} \tag{5}$$

The difference is about 1 min. Eq. (3) is less accurate not only because it uses an average value of pump speed, but because the values of S_o, S_n and P_a were read from a curve. The saving grace is that values given by Eq. (3) tend to be conservative. Eq. (5) is mathematically accurate, but has the disadvantage that R_{PS} and V_o are not always available in manufacturers' catalogs.

Part IV
Mechanical seals

A users' guide to mechanical seals
Call for higher quality is heeded by seal makers
When to select a sealless pump
Seals for abrasive slurries
Power consumption of double mechanical seals
Troubleshooting mechanical seals

Ivan Taylor and Bill Cameron, Flexibox Ltd.; Bill Wong, Bechtel Ltd.

A USERS' GUIDE TO MECHANICAL SEALS

If the engineer does not properly select, install and operate seals, they can pose serious problems.

Mechanical seals are devices that prevent leakage of a process fluid between a rotating shaft and a stationary housing by maintaining intimate contact between rotating and stationary seal faces. Thus, they are important and widespread throughout the chemical process industries, especially since they have replaced mechanical packings in numerous installations. This article provides basic working-knowledge that can aid engineers in seal selection and usage.

The main emphasis here is on what the seals consist of, how they operate, and what major kinds are available, along with practical tips on how to get the best out of a seal. For a look at trends in seal design and usage, see "Call for higher quality is heeded by seal makers," the following article. And for an explicit focus on improving service life, see "Troubleshooting mechanical seals" on p. 130.

What a seal consists of

In its simplest form, as shown in Fig. 1, a mechanical seal consists of five parts — the stationary face (1), a stationary secondary seal (2), the rotating face (3), a rotary secondary seal (4), and some device, in this case a single spring (5), to hold the faces together, to accommodate shaft misalignment and to transmit the shaft rotation to the rotary face.

The stationary face (also known as the stationary seal ring or stationary face, are suitable for the conditions and characteristics of the sealed fluid. The gap between the rotating face and the shaft is sealed by the rotary secondary seal.

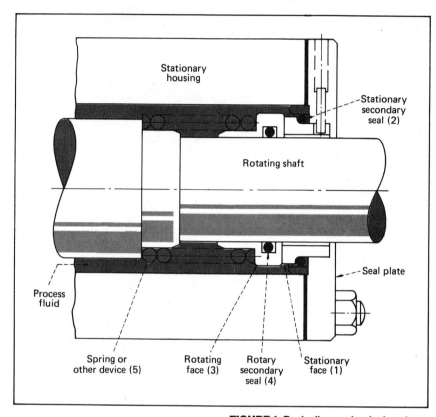

FIGURE 1. Basically, mechanical seals all have five key components. Many use a spring (above) for face closing.

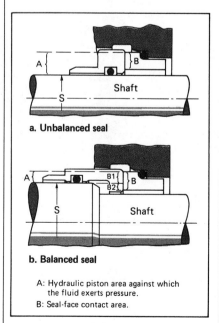

A: Hydraulic piston area against which the fluid exerts pressure.
B: Seal-face contact area.

FIGURE 2. Hydraulic balance depends upon hydraulic-piston and seal areas.

seat) is mounted in a seal plate (also known as a gland plate). This face is lapped to a flatness within approx. 0.0006 mm. and manufactured in materials compatible with the application. Different manufacturers argue the merits of using a stationary carbon face, to reduce chipping damage by mishandling, or having a stationary hard face, to reduce dynamic balance problems — in the final analysis, it's the face combination that counts.

The stationary secondary seal prevents leakage between the stationary face and the seal plate, and it may also provide resiliency for the mounting between the two. This seal may be a flat gasket, an O-ring or a rectangular section, made of elastomers, polytetrafluoroethylene (PTFE), graphite, or a proprietary composite.

The rotating seal face (rotary seal ring or seat) rotates with the shaft. This face is also lapped optically flat, and is manufactured in materials that, in combination with those of the sta-

Most variations in mechanical seal design are due to different face-closing mechanisms. In addition to keeping the seal faces together when the fluid is not pressurized, these devices can (1) provide the drive for the rotating face and (2) take up misalignment between the shaft and stuffing box housing while maintaining full contact between the seal faces. As regards their face-closing mechanism, there are two basic categories of seals: "pusher"-type seals, which employ one or more springs, and bellows-type seals.

The simple pusher-type design shown in Fig. 1 uses a robust, single coil spring that grips both the shaft and the rotary seal ring and transmits the drive without need for any additional mechanism. Other pusher-type arrangements use multiple springs. These smaller light springs are arranged in an annulus or pockets concentric with the shaft axis. Such seals need some other method of providing the drive; for that purpose, they typically employ pins, keys or dimples, although these can be subject to fretting problems.

The bellows seals employ axial bellows that envelop the shaft; they can be of metal, polytetrafluoroethylene, or an elastomer. In the metal bellows seal, the rotating face is flexibly connected to a drive ring, which is positively driven by the shaft. Metal and PTFE bellows seals provide the drive through the bellows itself (which can also take up misalignment). For elastomeric bellows seals, the drive relies on metal-to-metal contact of driving lugs, although a light spring may also be present.

In pusher-type seals, the rotating secondary seal is at the face end (see Fig. 1); in bellows-type seals, it is at the drive end, fitted between the shaft and the bellows drive ring. In the former case, the secondary seal must slide with the seal face to take up wear and axial displacement.

Apart from how pusher and bellows seals are designed, there are significant differences between them as regards operating characteristics and capabilities. These are summarized in Table I.

Seal assemblies also differ according to whether they are hydraulically balanced or unbalanced. The difference, very relevant to seal operation, is dis-

TABLE I. OPERATING CHARACTERISTICS OF PUSHER-TYPE AND METAL-BELLOWS SEALS

	Pusher-type seal		Metal-bellows seal
	Single-spring	Multispring	
Fluid-pressure limit	High	Very high	Medium
Fluid-temperature limit	Medium/high	Medium	Very high
Rotational-speed limit	High	Very high	High
Ability to handle suspended solids	No problem	Might accumulate in the small springs	Rotating bellows are self-cleaning; thin-section bellows are unsuitable for abrasive slurries
Ease of refurbishing	Easy as a rule	Varies with design	Ultrasonic cleaning may be needed
Ability to handle viscous fluids	Some designs need additional drive support to overcome viscous drag		Additional drive mechanism usually needed
Problem with seal hangup?	Occasional problem with crystallizing or otherwise solidifying products		No problem

TABLE II. POTENTIAL EFFECTS OF DUTY CONDITIONS ON MECHANICAL SEALS

Duty condition	Effects and considerations
High rotational speed	Distortion due to centrifugal stresses Vaporization due to high frictional heat at seal faces Dynamic imbalance if seal is very large Power losses due to churning, especially on rotating multispring seals Opening of single-spring coils if seal is very large
High speed plus misalignment	Fretting High wear rate for seal faces Fatigue in bellows seals
High fluid pressure	Distortion of seal faces Extrusion of secondary packings
Pressure X speed (PV)	The so-called PV factor is a guide to the wear resistance of the seal faces and, therefore, to the expected service life of the seal. Manufacturers specify a PV limit, to ensure that the seal delivers a good working life.
High fluid temperature	Limitations on material suitability Vaporization Carbonization of the product Creep in bellows, and in polytetrafluoroethylene components Alteration of shrink fits, due to dissimilar expansion coefficients Thermal distortion Safety problems
Low fluid temperature	Limitations on material suitability Increased viscosity, or solidification, of the product Deposition of dissolved solids Alteration of shrink fits, due to dissimilar expansion coefficients Thermal distortion
Variations in duty	Stop-start operation may prevent seal from settling into operational stability Pressure fluctuations can cause alternating concave and convex seal-face configurations, leading to excessive leakage and rapid seal-face wear Temperature excursions from the normal can lead not only to vaporization but also to alternating seal-face configurations, causing excessive leakage and rapid seal-face wear

cussed in a following section. Furthermore, mechanical seals are generally available in not only their normal configuration but also an easy-to-install cartridge configuration, wherein all the seal parts, including the seal plate, are built onto a common shaft sleeve.

How a seal works

An engineer charged with specifying and selecting a mechanical seal should not only be familiar with its components and assembly but also understand the principles whereby it works.

As noted earlier, the function of a seal is to prevent or minimize leakage between a stationary housing and a rotating shaft. We have seen that secondary packings prevent leakage between the rotary face and the shaft, and between the stationary face and the housing. So, the primary potential leakage path is between the rotary and the stationary faces, and these two components generally determine the limitations of the seal.

In performing its function, the mechanical seal maintains a thin film of liquid between stationary and rotating faces, which serves as a lubricant. Thus, a simple analogy is sometimes drawn between a mechanical seal and a plain thrust bearing. But whereas the bearing is lubricated by a plentiful flow of fluid selected in particular for its lubricating properties, the seal faces instead carry only a minimal amount of the sealed process fluid, which may well have poor lubricating properties. What's more, the copious flow of lubricant can readily remove heat at a bearing, whereas heat generated during seal operation must mostly dissipate through the seal parts themselves.

There are many complex theories as to how the faces of a seal are kept lubricated, but suffice it here to say that a liquid film is maintained between the faces by a combination of fluid pressure and the relative rotary motion of the faces. It is the gap between the faces, and the condition of the fluid therein, that determine whether a seal is working correctly: Too large a gap and the seal leaks; too small a gap and the faces run dry. The ideal situation is to have just sufficient fluid between the faces so that most of the system is running on liquid but the portion of fluid exposed to the atmosphere is receiv-

TABLE III. HOW SEALS COPE WITH CHARACTERISTICS OF THE SEALED PROCESS-FLUID

Process-fluid characteristic	Courses of action
Corrosivity, chemical reactivity	Select appropriate seal materials Provide seal faces with a flush by compatible liquid Provide liquid barrier by using double or tandem seal
Dissolved solids	Provide solvent or steam quench outside seal, to prevent solidification Employ two hard seal faces (e.g., silicon carbide against tungsten carbide) for products that tend to crystallize Provide seal faces with a flush by compatible liquid Use double or tandem seals
Suspended solids	Make sure the fluid reaching the seal faces is solids-free; use cyclone separator or magnetic filter if appropriate Employ two hard seal faces (e.g., silicon carbide against tungsten carbide) for products with abrasive particles Employ external-spring design, to keep solids from accumulating around the spring and to maintain a clean region under the secondary sliding packing Keep in mind that with multispring designs, solids tend to accumulate in multispring pockets
High viscosity; solidification	Build in additional torque-carrying capacity to overcome viscous drag Preheat the product, and the pump, to overcome high-torque startups Circulate a solvent before operation, to remove tars and similar materials Employ steam jacketing and trace heating, to keep viscosity down Use large-diameter stuffing boxes, to allow freer circulation of the product over the seal
Partial vaporization from mixtures; presence of dissolved gases	Remember that control of vaporization is more difficult when the fluid is a mixture of volatile components, because its vapor pressure may be higher than that of the lightest component Keep in mind that the heat generated at seal faces can drive dissolved gases out of the solution Consider symmetrical cooling of the seal faces by using a multipoint injection collar, to avoid localized hot spots
Explosivity; toxicity; other hazardous properties; high value	Bear in mind that safety of personnel and equipment may require security measures tailor-made for the situation Consider a close-clearance safety bushing, to prevent rapid loss of fluid to the atmosphere if the seal fails Consider a lip seal and a safety bushing, to retain an external flushing fluid Consider a backup dry-running seal for emergency operation Consider a double-seal system (pressure of the barrier liquid is higher than that of the sealed product, so any leakage across the inner seal will flow into the product) Consider a tandem-seal system (pressure of the barrier liquid is lower than that of the sealed product, so any leakage across the inner seal will flow into the barrier liquid)

ing sufficient heat to turn it to vapor. Under these conditions, no visible leakage will be seen and a good seal life should ensue.

Problems arise when the greater part of the very thin film breaks down and starts to boil. This condition, known as vaporization, can be recognized by either loud popping sounds coming from the seal or puffs of vapor coming from between the faces. The seal will fail quickly if vaporization is not rectified. Preventing this vaporization is a major goal during seal selection.

Balanced, unbalanced

A key factor in seal operation is the hydraulic balance principle, which is relevant for maintaining a stable liquid film at the seal faces under the specified operating conditions. A given seal assembly is either balanced or unbalanced, depending on the relationship between the pressure of the fluid being sealed and the pressure of contact between the seal faces.

In the unbalanced arrangement, Fig. 2a, all of the seal-face contact area lies outside of the shaft diameter S, and the hydraulic piston area (A) against which the fluid exerts presssure is greater than the seal-face contact area (B). This seal has a balance ratio A/B greater than 100%, as do all unbalanced seals. On the other hand, in a balanced seal, Fig. 2b, only part of the contact area (B_1) lies outside of that shaft diameter; the rest of the contact area (B_2) is inside it, because of a step (a narrowing of the diameter) in the shaft. Since B_1 is equal in area to A, the balance ratio $A/(B_1 + B_2)$ (which is equal to A/B) is less than 100%. For most balanced seals, this ratio lies between 0.67 and 0.8.

Because balanced seals enable seal-face contact pressure to be kept below the sealed-fluid pressure, a thicker film of stable liquid can exist between the faces, reducing friction and the consequent heat generation. Thus, a balanced seal can handle more-arduous, higher-pressure duties than can an unbalanced one. (The heat generated by friction at the seal faces can be reduced even further by using a narrower seal face on either the rotary or the stationary seal ring.) Normally, a balanced seal is designed to operate with the lowest face pressure that will effectively prevent leakage between the faces.

Selecting the right seal

The style of seal used will depend largely on the process and on the other equipment — particularly the pumps — for the project. In the early stages of a project, the process units and equipment (particularly pump) duties will be listed and categorized. A decision can then be made as to which general seal types will cover the majority of duties.

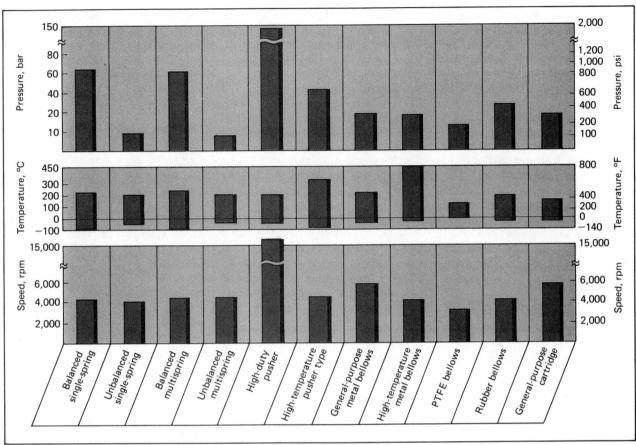

FIGURE 3. Numerous seal types can accommodate various ranges of fluid pressure and temperature and shaft rotational speed.

Next, a more-detailed review will lead to the choosing of more-specific types of seals, together with ancillary services required to enable them to operate correctly. Any appropriate API seal codes should be brought into the picture.

By this stage, the engineers specifying the seal will have taken into account such operating factors as process-fluid temperature and pressure at the seal, speed of rotation, and shaft size. A detailed summary of such major duty-conditions and their effect on seals appears in Table II.

Equally important are the chemical and physical characteristics of the process fluid being sealed. How mechanical seals cope with various fluid characteristics is summarized in Table III.

In addition to meeting the operating specifications, engineers involved in seal selection often have to consider other factors, which may bear no direct relation to the sealing function. A key example is environmental regulation that imposes strict leakage criteria. One way of meeting these is to use double seals, described further on in this article.

A multitude of versions

Mechanical seals come in a multitude of designs and sizes. Different manufacturers use different features to achieve the sealing function and may specify different operation limits. A guide to the operational capabilities of numerous major types appears in Fig. 3. However, because of the variety of seals available from many manufacturers around the world, Fig. 3 is not exhaustive. Here are comments about the various types:

• *Single-spring pusher seal*: Available in balanced and unbalanced designs for most process duties, this is the simplest mechanical seal. The single-spring feature reduces the possibility of spring choking by solids, such as crystals or catalyst fines. Variations available include dual-rotation, short-length, stationary-spring and external-spring. The last-named (Fig. 4) extends the seal size and speed range; since its functional spring is located outside the sealed process fluid, the spring is not open to chemical attack or clogging.

• *Multispring pusher seal*: One of the most common mechanical seals used in chemical-process use, particularly with pumps. Secondary sealing between the seal face and the rotating shaft is provided by an elastomeric component, O-ring, wedge, or U-cup section ring acting under a small amount of compression. The multispring arrangement is available in balanced and unbalanced configurations, yielding a bidirectional seal that is usually shorter than the single-spring version. Frequently, multispring seals are used in pairs to produce simple double-seal arrangements (Fig. 5), employing two single seals back-to-back with an innocuous barrier fluid in-between; these are especially attractive when process-fluid leakage is completely unacceptable, whether for economic, environmental or safety reasons. A related arrangement of tandem seals is also used.

• *Modified pusher seal for severe service*: Design features include stationary springs, anti-distortion components and dynamic balance, resulting in a seal for high pressures and high speeds.

• *General-purpose metal-bellows seal*: This version was developed to overcome hang-up problems associated with the sliding secondary seal of the pusher-type seals. Although the failure mode tends to be catastrophic if the bellows fails, whereas a pusher-type seal normally deteriorates progressively, metal-bellows seals are gaining popularity in the process industries.

• *High-temperature metal-bellows seal*: Selection of materials with similar expansion coefficients, and replacement of the conventional elastomeric secondary seal by high-temperature material, usually an exfoliated graphite, allows the bellows seal to operate at temperatures up to 450° C.

• *Modified metal-bellows seal*: Designed for high-temperature and moderate-pressure applications. Usually featuring a laminated bellows, it extends the pressure range of the metal-bellows seal to 50 bar.

• *Nonmetallic-bellows seal*: The PTFE-bellows seal is mounted externally, so that the minimum number of parts are wetted by the sealed product. PTFE-bellows seals have wide application for chemically aggressive liquids in pumps and mixers. In the "rubber bellows" seal, the faces are held together by a nondriving spring. This is a low-pressure seal found in offsite water systems, and widely used in the household and automobile markets.

• *General-purpose cartridge seal*: As all users have become more aware of the costs of maintenance, seal makers have developed general-purpose cartridge designs. The aims are to reduce inventories by offering one seal for a variety of applications, and to minimize downtime. Many single-spring, multispring and metal-bellows seals are already available in cartridge designs, and more are coming.

The best from your seal

Most chemical-process applications for mechanical seals involve pumps. Before pumps are shipped, their performance is often checked by factory test — which provides an opportunity to likewise check the integrity of the seal. At this stage, the suitability of the seal for the physical conditions can also be established, so long as the fluids used during the test are compatible with the seal materials selected for the final duty conditions. In the case of double or tandem seals, the pump-testing also offers a chance to check out the barrier-fluid system.

It is, of course, important to ensure that the pump and associated seals arrive onsite in as good a condition as when they left the factory. Typical shipping-protection techniques include fitting blank plates to nozzles and flanges, plugging any tapped holes, and injecting the internals and coating the externals with suitable inhibitors. These measures keep out dirt and forestall corrosion; either could result in debris circulating through the pump during startup, causing damage to seals and other close-clearance items.

Occasionally, special treatment is given to the seal itself. It can be removed, shipped separately and reinstalled onsite, or it can be stripped from the pump onsite, cleaned and reinserted. This latter approach obviously presents less chance than does the former for debris to enter the pump.

Seal life is greatly influenced by factors outside of the operating environment of the seal itself. For example, heavy vibration, misalignment of the pump casing and shaft, bent shafts and worn bearings all affect the operating life because of the vibrations or fatigue motion they impart to the seal components. Great care should be taken in assessing bearing loads, lining up pump shafts and gland plates, and, especially, aligning the driving and the driven machinery. Bearing failures will inevitably result in seal failures, and misalign-

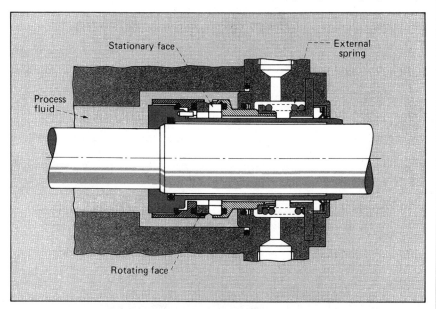

FIGURE 4. In some types of pusher seals, the face-closing spring is external. Since it is thus not in contact with the process fluid, the fluid cannot attack or clog it.

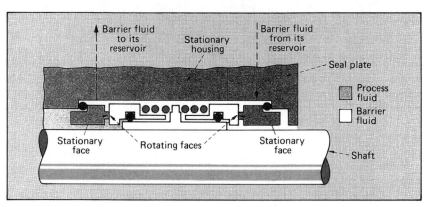

FIGURE 5. In double seals, harmless barrier fluid keeps the process fluid from escaping to the outside. Also, barrier fluid leaks into process fluid, not vice-versa.

ment-induced vibrations will cause rapid seal wear, either to the faces or to the secondary packings.

Fully-instrumented process monitoring can keep track of the status of seals, and allow planned, rather than emergency, maintenance. The future will see more use of these units, as cost savings favor unspared machinery and remote monitoring.

Other applications, too

Although mechanical seals are associated especially with rotating pumps, they are also widely used for sealing other rotating shafts: on, for example, agitators, mixers, reactors, scrapers, compressors and fans. Except for the use with scrapers and some of the lower-duty mixers and agitators, these applications require specialist knowledge of the machinery and process for the design of the best sealing solution.

Indeed, such seals may be unique items, designed especially for the particular machine. For example, seals for centrifuges can exceed 8 in. (200 mm) dia, while also requiring chemical- and heat- resistant materials such as Cabot Corp.'s Hastelloy C or Du Pont's Kalrez. Although such special-purpose machinery more often than not rotates slowly, it nevertheless can require special considerations such as dry running, split seal faces for ease of maintenance, or a very large seal diameter. In all of these cases, consultation with the seal maker is prudent.

We would like to acknowledge the help and support of Phil Ansley-Watson of Bechtel Ltd. in the preparation of this article.

The authors

Ivan Taylor is group marketing manager for Flexibox International, Nash Rd., Trafford Park, Manchester M17 1SS England. After 14 years with Rolls-Royce, he moved to Flexibox in 1978, where his other positions have been engineering manager and international contracts manager. He has an honors degree in mechanical engineering, is a chartered engineer, and is a member of both the U.K. Institution of Mechanical Engineers and the American Soc. of Mechanical Engineers.

Bill Cameron has been a technical author within Flexibox's U.K. Marketing Dept since 1984. His previous positions with the company include sales engineer, senior export engineer, and product support engineer. He has a national diploma in mechanical engineering with a marine engineering endorsement, plus a national certificate in business studies. Before joining Flexibox, he was with Ocean Fleets Ltd., Liverpool.

Bill Wong, chief mechanical engineer at Bechtel Ltd., Bechtel House, 254 Hammersmith Rd., London W6 8DP, England, has had over 30 years experience in the manufacture, development, application and operation of a wide range of rotating equipment. Before joining Bechtel in 1974, he worked with Dresser Industries, Cooper Energy Services, and Air Products & Chemicals Inc. He obtained his technical education in London, and has written about rotating equipment for various organizations.

Call for higher quality is heeded by seal makers

Manufacturers and clients alike are working to reduce the mean time between failures. Seal reliability also will be enhanced by new standards that allow better performance.

Prodded by a demand for longer-lasting seals, equipment manufacturers are introducing a variety of improved designs, and the prospects are that product reliability will get much better in the future.

One factor in favor of improved reliability is that both the American National Standards Institute (ANSI) and the American Petroleum Institute (API) plan to introduce modified centrifugal-pump standards that include more stuffing-box space. This, say industry spokesmen, will permit the use of larger seals, and extend seal life because of better heat dissipation, lubrication, and more-efficient dirt removal.

To be sure, seal buyers seem to be increasingly reluctant to cut corners on quality, observe manufacturers, noting that sales of more-reliable cartridge and bellows seals, despite the higher cost, have been rising.

"The main thing that everybody is interested in is to reduce the mean time between failures (MTBF) and save money," says John Reynolds, chairman of the ANSI/ASME (American Soc. of Mechanical Engineers) B73 committee (the B73 standard covers chemical-process pumps).

TYPICAL PROBLEMS — While the need to avoid hazardous leaks has boosted sales of seal-less pumps (See "When to select a sealless pump," p. 118) meeting emissions standards is not generally a problem for manufacturers. Under the Clean Air Act's New Source Performance Standards and standards for certain hazardous pollutants, seals must be inspected monthly and repaired if the ambient organic concentration at the interface is 10,000 ppm or more. Process plants can even avoid this chore by using properly specified double seals, which have a barrier fluid between the primary and backup seals.

"Seals don't usually have much leakage unless they are wearing out or not properly installed," says K. C. Hustvedt, an environmental engineer with the chemicals and petroleum branch in the Environmental Protection Agency's Emission Standards and Engineering Div. (Research Triangle Park, N.C.).

Exxon Chemical Co. (Baytown, Tex.), which has studied pumps problems intensively, has estimated that the average MTBF for seals on ANSI pumps is 1.2 years, and the average cost of a failure is around $5,000 (See "Downtime prompts upgrading of centrifugal pumps," p. 23). Reynolds disagrees with Exxon's figures (see p. 30); other industry spokesmen feel that an MTBF of about 1 year is typical.

A general goal is to try to double the MTBF, says William Adams, vice-president, engineering and research for Durametallic Corp. (Kalamazoo, Mich.), a leading seal manufacturer. Adams, a member of the ANSI B73 committee, feels that the pump modifications will be an important step toward that goal.

ROOM FOR IMPROVEMENT — "The problem with the stuffing-box standards is that they were designed for packing, and are generally inadequate for mechanical seals," says Roger Peck, manager of marketing services with Goulds Pumps, Inc. (Seneca Falls, N.Y.). Peck estimates that 60–70% of the process pumps sold today have mechanical seals. "In the early 1970s, it was the other way round," he notes, adding that the popularity of mechanical seals continues to increase, though at a slower rate. He indicates that packing is still commonly employed in particular industries, such as pulp and paper.

The ANSI/ASME B73 committee proposes to retain the existing stuffing-box standard, but add a second option — called a "seal chamber" — with larger dimensions. The seal-chamber radii for the three pump-size groups in

Split seal can withstand a pressure of up to 150 psig.

Originally published May 12, 1986

CALL FOR HIGHER QUALITY IS HEEDED BY SEAL MAKERS

the B73 standard are 3/4 in. (compared with 5/16 in. for the stuffing box), 7/8 in. (3/8 in.) and 1 in. (7/16 in.) The working group studied only horizontal pumps, but vertical pumps will also be included. The changes are among several planned for the B73 standard; Reynolds says the revised version should be published sometime in 1987.

API's 610 standard, for heavy-duty centrifugal pumps for refinery service, is being revised to specify a minimum stuffing-box size, with the additional requirement that mechanical seals must have a minimum clearance of 1/8 in. to avoid overheating due to shear. The latter is a major cause of premature failures, says Frank Landon, chairman of the taskforce working out the API-610 amendments.

Expected to become effective in 1987, these changes include the following provisions for stuffing boxes: for shaft sizes up to 2 in., a minimum radial clearance of 1 in. between the shaft and the bore of the stuffing box, and minimum length of 5 3/4 in.; for shafts up to 3 in., minimum clearance of 1 1/8 in. and lengths of 6 1/2 in.; and for shafts larger than 3 in., a minimum clearance of 1 1/4 in., and length of 7 in.

"Up to now there has been no standardization," says Landon, who is chief of mechanical engineering and design with Fluor Corp.'s Advanced Technology Div. (Irvine, Calif.). "This will require about 30% of the manufacturers to redesign their stuffing boxes."

ACTING IN ADVANCE — Some equipment makers have decided not to wait for the new standards. A.W. Chesterton Co. (Stoneham, Mass.) long ago became impatient with the space allotted to pump seals and developed its own pump design, built around oversized seals (*Chem. Eng.*, Apr. 1, 1985, p. 14).

More recently, some pump manufacturers have introduced "upgraded medium-duty process pumps," whose enhanced features include larger mechanical seals and a magnetic seal in place of the conventional lip seal for the pump bearings. This was in response to a request of Exxon Chemical: "We didn't underwrite it, we simply asked various manufacturers if they could build a pump with these upgraded features," says Heinz Bloch, a senior engineering associate with Exxon.

Carver Pump Co. (Muscatine, Iowa), Goulds, and Wilson-Snyder Pumps (Garland, Tex.) are now making upgraded pumps. Although more expensive than conventional pumps, they are expected to last longer. Carver anticipates an average MTBF for its pump (not just mechanical seals) of 25 months.

Goulds' Peck says his firm's new pumps were tested "in very demanding service" in a chemical-process plant for a year, whereas previous units had lasted only about two months. He notes that, after mechanical seals, lip seals are the second most common cause of pump failure.

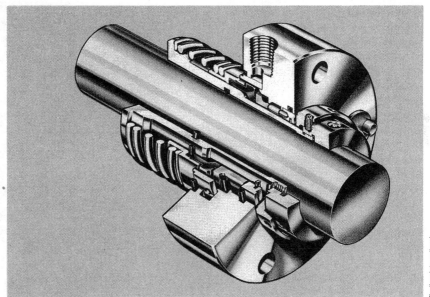

System for light hydrocarbons has a primary seal and a backup

Magnetic seals are used extensively in the aerospace industry; while expensive, they have operated as long as 40,000 h without repair, according to Exxon. (See "Downtime prompts upgrading of centrifugal pumps," p. 23.)

Exxon Research and Engineering Co. (Florham Park, N.J.) has also been pushing for seal improvements. The company has had its own seal research and development program for about six years "because we weren't able to get the kind of reliability from vendors that we needed for our process pumps," says Thomas Will, a senior staff engineer. "Our idea wasn't to make our own seals, but just to improve the technology, then have the vendors meet our standards."

As a result, Exxon now has its own specifications, and requires suppliers to meet these if the seals are not readily available.

GAINING ACCEPTANCE — Cartridge seals are becoming increasingly popular, say industry spokesmen. Although they cost roughly twice as much as conventional types, they can be installed quickly and easily, and they avoid the potential problems of faulty installation. Reynolds feels that a larger seal chamber on pumps will further boost cartridge sales by giving designers more flexibility.

Metal bellows seals, whose accordion-like design and action eliminate the plugging that can occur with spring designs, are also gaining rapidly in popularity, though they can cost up to 50% more.

"Five years ago, we were about the

only company making bellows seals, but lately everybody has been getting into the business," says Steven Palmer, director of marketing information for EG&G Sealol's Process Sealing Div. (Cranston, R.I.), which claims to be the leading manufacturer of such units.

Double seals, which have a secondary seal, are increasingly in demand to reduce leakage risks, he says. Aside from having a backup seal, they are cooled by a buffer fluid that reduces the potential for failure in high-temperature operations.

An advantage of double-bellows types over conventional double units, adds Palmer, is that the secondary seal does not move, so there is no danger of shaft wear. Last year, Sealol came out with what it claims is the first tandem arrangement in a double-bellows seal — a design said to offer more protection in case of primary seal failure.

John Crane-Houdaille, Inc. (Morton Grove, Ill.) introduced at the end of March a series of what it calls second-generation metal bellows seals. The bellows leaf has a "double-nested" design, with the angle of the convolution tilted to transfer stress concentration away from the welds, which are at the outer and inner edges of the leaves. The area around welds has the greatest failure potential, explains Richard Tempinson, a marketing manager. Crane offers a two-year warranty on the seals, whereas the usual is about one year (bellows seals are normally used in hot, harsh environments).

The new design "results in a bellows that can take higher temperatures and greater variations in pressure," claims George McLaughlin, vice-president of marketing and sales. Depending on the types and materials specified, temperature ranges for the new seals can be as great as –100°F to +800°F, with recommended pressures up to 500 psig, the company says. The new seals are also available in cartridge form.

Crane Packing, Ltd. (Cheltenham, England), an affiliate of John Crane, also has a new bellows seal (not yet available in the U.S.) said to distribute stress more evenly than do conventional types. It has an asymmetric design — the radii at the crests of the bellows' convolutions are greater than those at the roots of the convolutions. Also, it has only 7 components, vs. a typical 34, and this is said to reduce the risk of failure and the cost. Crane Packing's new seal is specified for temperatures of –40°C to +200°C, and pressures to 20 bar.

Other companies also make high-temperature bellows seals. Last year, Garlock Inc. (Palmyra, N.Y.) brought out an 800°F-rated model.

And various companies are manufacturing cartridge bellows seals. In 1985, Durametallic put one on the market; it has a bidirectional lubricant-pumping feature, and can be installed on either end of a double-ended pump. The company has applied for a patent on the pumping mechanism — a vane built into the gland to circulate the buffer fluid.

KEEPING IT CLEAN — Dry-running seals are rapidly increasing in popularity, especially for use in mixers and compressors, where product contamination from seal-buffer fluid may be a problem. The units use face combinations of carbon graphite with tungsten carbide or ceramic. Face geometries are specially designed, and the faces carefully balanced to dissipate heat.

Crane makes dry-running compressor seals that can operate up to 1,500 psi at surface speeds up to 500 ft/s, says Joseph Sedy, technical director of advanced applications engineering. The seals substitute for conventional high-pressure oil-lubricated ones between the compressor and its bearings (labyrinth seals are used between compressor stages), and employ the compressor gas instead of an oil lubricant. This not only avoids oil contamination, but also saves about $500,000 in the cost of the oil seal system.

"Conversion to our system costs $30,000 to $80,000," says Sedy, "and the annual savings in operating costs can be as high as $100,000." Crane installed a handful of test systems between 1977 and 1982, he says, "and in the last two years we have been doing a conversion about every two weeks." Since gas-operated seals are non-contacting, they are said to last longer than conventional models. Crane recently brought out a new series for a variety of applications, and anticipates lifetimes of 5-15 years.

Leakage rates from seal buffer fluid in mixers can be high because of the pressure cycles and because the shafts are long and tend to have a lot of deflection, says Durametallic's Adams, who

Made-in-Britain rotary seal is being tried in pump retrofits

notes that the company's sales of dry-running seals increased 100% last year and 50% the previous year.

Mixing Equipment Co. (Rochester, N.Y.) has experienced a steady increase in products that use dry-running seals, and is working on new designs, says David Mechler, director of sales. One of the current designs has a metal "dry well," which is a metal plate that pre-

CALL FOR HIGHER QUALITY IS HEEDED BY SEAL MAKERS

vents materials from the seal from falling into the tank when the seal is flushed.

Mechler estimates that about two thirds of mixers that require sealing use packing and the rest use mechanical seals, compared with three quarters that employed packing 10 years ago, "about the time EPA and OSHA started clamping down on emissions." Both Mechler and a spokesman for Chemineer Inc. (Dayton, Ohio) say many clients prefer packing because the initial cost is low and it is easy to work with, but that mechanical seals are easier to maintain and may be cheaper in the long run, depending on the application.

AVOIDING DAMAGE — A common problem with mixer applications is that seals may be damaged by shaft deflections. Manufacturers try to minimize the risk by putting the machine's second bearing (or a third bearing) on top of the stuffing box, near the seal, says Michael Habich, a senior design engineer with Chesterton's Mechanical Seal Div. (Groveland, Mass.).

Chesterton's answer to this problem is a double, cartridge-mounted seal that has increased clearance between the rotating and stationary parts to allow for deflections.

Although Garlock offers an expansion joint option (it goes on top of the mixer and absorbs movement), it is working on what it thinks may be an even better solution. "We are developing one that spring-loads the [outer] stationary seal instead of the rotating seal," says Timothy Donovan, manager of marketing for mechanical seals. "We think it will have better compensation for misalignment."

BETTER MATERIALS — Tungsten carbide is the most popular material for seal faces, but the use of silicon carbide is increasing (although it is more fragile), mainly for high-temperature applications. The seal industry is seeking better ceramics formulations.

"What's needed is a material that has good thermal conductivity and abrasion resistance, like silicon carbide, but is more impact resistant," says Exxon's Will. Designers are also working on ways to prevent seal faces from distortion under stress, basically by designing the units so that distortion is controlled.

BEING SPECIFIC — Some companies have come up with radically new designs to handle conventional sealing jobs, or have designed seals for specific, difficult problems.

Utex Industries, Inc. (Houston), a manufacturer of molded seals and gaskets, has developed a device called Ezeseal as a substitute for a mechanical seal. Ezeseal is a rotating housing assembly that is attached to the pump shaft. Leakage along the shaft is propelled to the periphery of the housing assembly by centrifugal force. There, it is picked up by a pitot-tube system and returned to the stuffing box. The advantages, says Utex, include less down-

> *Split mechanical seals, which are assembled in two halves around the shaft, have the advantage that the pump or other equipment does not have to be disassembled.*

time and longer life, because there are no critical wearing parts (for more details, see "Trouble shooting mechanical seals," p. 130).

Flexibox International (Manchester, England, and Houston, Tex.) has developed a sealing system with a primary and secondary (backup) seal for light hydrocarbons, which are not only difficult to seal effectively, but also potentially flammable or explosive. The system, rated at up to 40 bar, was put on trial at Esso Petroleum's refinery at Fawley, near Southampton, England, in February 1984.

As the shaft spins, a pressure drop is created between the seal's inner and outer edges. At the same time, ports in the rubber ring admit pressurized fluid into the seal's radial cavity, squeezing the seal inward and progressively reducing the gap between the sleeve and shaft until the seal is operating on a fluid film only a few micrometers thick.

Another British company, James Walker & Co. (Woking), has come up with a novel rotary seal that clients are trying as retrofits for pumps. At this point, the unit is limited to applications below 180°C by its elastomer and PTFE materials of construction. It also cannot tolerate situations in which there is a phase change — e.g., from liquid to vapor. The company is working to solve both limitations.

COSTLY ASSEMBLAGE — Split mechanical seals, which are assembled in two halves around the shaft, have the advantage that the pump or other equipment does not have to be disassembled. However, they tend to be about triple the cost of conventional seals because they are difficult to make (the whole pieces must be carefully broken or cut and lapped so that the surfaces will match tightly). Also, assembly is a rather delicate operation, and the units are generally limited to low-pressure applications.

"They are typically limited to 25 psi," says Henri Azibert, a design engineer with Chesterton's Mechanical Seal Div. "When the pressure goes higher they start to fall apart, although you can get some very expensive ones that can handle up to 100 psi."

Chesterton claims to have set a precedent with the introduction earlier this year of split seals that cost about 20% less than its standard single-cartridge seals and can handle up to 150 psig ("we plan to go to higher pressures," says Azibert.) The key is that, because of their geometry, the seals are pressurized from the outside, whereas conventionally split seals are internally pressurized and require heavy clamps. Azibert notes that the outer, stationary seal is located inside the gland plate, which is outside the stuffing box. This provides more space for the assembly.

A seal that is essentially a thick rubber ring and uses standard face materials is available from the Mechanical Seal Div. of Borg-Warner Industrial Products, Inc. (Long Beach, Calif.). The company calls it a "rubber-in-shear" seal. Developed for slurry pumps, it does not use springs, but is convoluted, and looks somewhat like a bellows type. A company spokesman says that it has been employed successfully in alumina plants for a couple of years, and is being field tested in the pumping of slurries of gold, nickel, copper, limestone and coal. Unlike conventional packing or seals, it does not need flushing with water. This, claims the firm, saves $8,000–$15,000/yr per pump by eliminating the need to evaporate additional water from the product.

WHEN TO SELECT A SEALLESS PUMP

Sealless pumps prevent leaks and prolong the pumping time between maintenance calls.

Ali M. Nasr, Liquiflo Equipment Co.

Sealless pumps are often the preferred choice for pumping hazardous and toxic fluids, because they are designed not to leak. They also have an economic advantage in certain less-critical services, though they may cost more initially than conventional pumps: Sealless pumps require less maintenance, and are less likely to fail and cause expensive downtime, because they have no mechanical seals.

This article discusses the capabilities, costs and benefits of five types of sealless pumps: magnetically-coupled gear pumps and centrifugal pumps, canned-motor centrifugal pumps, hydraulically-backed double-diaphragm pumps, and pneumatic-powered diaphragm pumps.

Sealed and sealless pumps

A conventional pump has a mechanical seal around the driveshaft to prevent liquid within the pump from escaping into the environment. Unfortunately, all mechanical seals leak to some slight extent. As they wear down, or get corroded by the process fluid, they leak more and more, and eventually they fail. For this reason, all sealed pumps have to be torn down regularly so that the seals can be repaired or replaced. (Leakage is not such a problem with double mechanical seals that can be flushed continuously, though it may be difficult to dispose of the flushed liquid.)

Sealless pumps are designed not to leak at all. For this reason, they are very widely used for handling toxic, noxious and explosive liquids — e.g., acids, bleaches, poisonous solvents, and active pharmaceutical compounds. In other cases, they can often be justified on purely economic grounds: Sealless pumps require less maintenance than sealed pumps, which saves on maintenance cost and reduces downtime. They are also less likely to cause spills.

Fig. 1 shows the workings of four types of sealless pumps. In the gear and centrifugal types, note that the pumped fluid is isolated from the atmosphere by solid metal. In the double-diaphragm type, the isolation is achieved by two flexible membranes. Let us explain each type in detail.

Sealless centrifugal pumps

Centrifugal pumps are the workhorses of the chemical process industries (CPI) because of their durability, efficiency and range of materials. In a sealed design, the seals and shaft are the only wearing parts. In a sealless design, the manufacturer replaces the standard shaft and seal with a sealless drive.

Fig. 1a shows a magnetically-coupled centrifugal pump. The impeller shaft, surrounded by magnets, is supported on bearings within a metal containment can. Outside the can is another set of magnets, attached to the motor shaft. As the motor turns the outer coupling, magnetic attraction forces the inner coupling, and thus the shaft and impeller, to turn also.

Magnetically-coupled centrifugal pumps are as efficient as standard ones. What limits them is the amount of torque the coupling can transmit. This depends on the magnets used: A typical 1.4-in.-long ceramic coupling transmits about 33 in.-lb of torque; a samarium-cobalt coupling transmits 83 in.-lb; and the newest type, neodymium-iron, transmits 110 in-lb.

Manufacturers typically provide a torque rating for a particular temperature — torque capability declines as temperature increases. As for capacity and size, magnetically-coupled centrifugals range from about 1 gpm to 300 gpm, and ¼-in. to 3 in. Most such pumps are limited to 100–200 ft of head: In

Originally published May 26, 1986

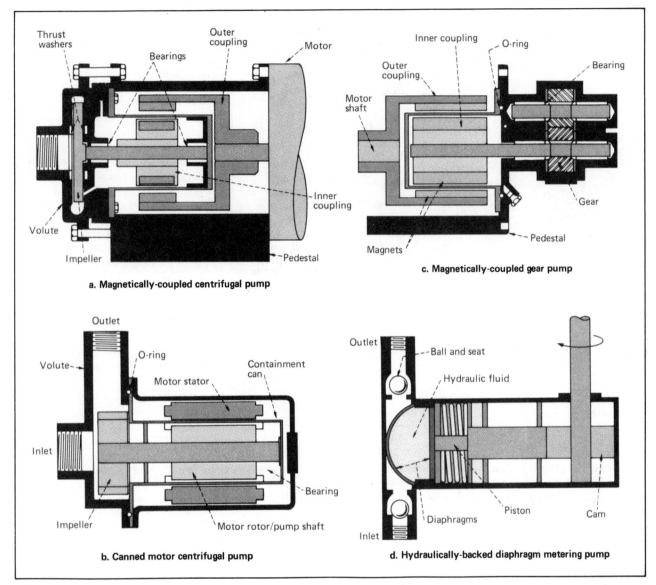

Figure 1 — Sealless pumps use magnetic couplings, canned motors, or diaphragms to isolate pumped liquid (yellow) from atmosphere

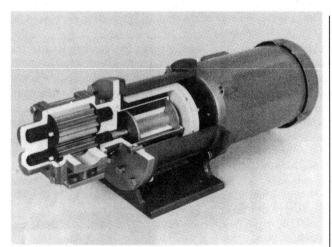

Figure 2 — Cutaway of magnetically-coupled gear pump; inner coupling is also cut away

higher-head applications, the thrust washers are prone to wear, and so gear pumps are usually the better choice.

The canned-motor centrifugal, shown in Fig. 1b, is the other sealless type. Capacities range from about 10 gpm to 300 gpm. In this design, an electric-motor rotor sheathed in metal serves as the impeller shaft. It rests on bearings inside a containment can; the motor stator is outside the can.

Such pumps are built by machining an electric-motor stator and rotor to make room for the shaft sleeve and containment can. Since this alters the standard dimensions, it makes for a less efficient motor that gives off more heat. For this reason, such pumps often include cooling systems. Magnetically-coupled pumps do not have such heat buildup.

Sealless gear pumps

Magnetically-coupled gear pumps can achieve differential pressures of 100–150 psi, and discharge pressures of 300–350 psi, and are available in capacities from less than 1 gpm to

120 MECHANICAL SEALS

Table I — Sealless pumps tend to cost more initially; centrifugals are the most efficient

a. Approximate initial cost, including drive and base assembly

Capacity, gpm	Centrifugal Plastic	Centrifugal 316 SS	Gear 316 SS	Diaphragm Metering	Diaphragm Other
1	N.A.	N.A.	$200–250	$400–1,000	N.A.
1–3	$100–200	N.A.	400–1,200	600–1,000	N.A.
10	300–500	$1,100	1,500	6,000–10,000	$360–1,000
20	500–800	1,100–1,500	1,700	N.A.	460–1,500
30–50	1,400	3,000–6,000	3,500	N.A.	2,000
100	3,000	4,000–6,000	N.A.	N.A.	3,750

b. Power cost for sealless 20-gpm pumps (70-ft head).

Type of pump	Initial cost	Annual power cost	Total 5-yr cost
Centrifugal, plastic	$650	$225–675	$1,775–4,025
Centrifugal, 316 SS	1,300	225–675	2,425–4,675
Gear, 316 SS	1,700	350–1,050	3,450–6,950
Diaphragm, plastic	450	790–2,370	4,400–12,300
Diaphragm, 316 SS	1,500	790–2,370	5,450–13,350

Basis: 6¢/kWh, 8–24 h/d.

c. Approximate initial cost of sealless vs. sealed pumps.

Capacity, gpm	Centrifugal Sealless	Centrifugal Sealed	Gear Sealless	Gear Sealed
1–3	$100–200	N.A.	$400–1,200	$650–900
10–20	300–1,500	$450–600	1,500–2,200	1,000–1,200
30–50	1,100–1,500	550–800	3,500	2,500
100	3,000–6,000	1,800–3,000	N.A.	N.A.

N.A.—Not available

about 55 gpm. Fig. 1c shows their design; Fig. 2 shows an actual pump in cutaway. Such pumps can handle fluids to about 8,000-cP viscosity; their sealed counterparts, which have higher torque capabilities, can handle 100,000 cP.

Sealless gear pumps are more prone to wear than centrifugals because they have more bearings, plus internal seals. Such pumps are used to transfer and circulate difficult fluids — they have been known to run for more than a year without maintenance in H_2O_2 and SO_3 service — and for high-head, high-viscosity and suction-lift applications.

Gear pumps require more torque than centrifugals, and have to be capable of handling the startup torque as well as the running torque. The startup torque is about twice as great because the fluid in the line has to be accelerated. Such torque requirements once limited gear pumps to capacities below 10 gpm, but more-powerful magnetic materials have extended their capacities to 50 gpm or more.

Diaphragm pumps

These pumps never have seals; the design does not require them. Moreover, they are generally selected not for their sealless design but for their ability to meter precisely and to handle abrasives, slurries and highly viscous liquids.

Diaphragm pumps (Fig. 1d) are available in capacities from about 3 gal/h to 1,000 gpm or more. Pneumatic diaphragm pumps are typically limited to 100 psi differential because they are powered by air at around that pressure. Mechanical and hydraulic pumps can develop several hundred psi; 600 psi is not uncommon for small metering pumps.

Because diaphragm pumps have a pulsing action, they cannot be used in applications that demand steady flow. And because they are less efficient than centrifugals or rotary pumps, they are used mainly for fluids that other pumps cannot handle.

How sealless pumps fail

If properly designed, and built with the right materials, sealless pumps generally fail less often than sealed pumps. But they do fail, and in order to prevent failure it is important to understand how it can happen.

In magnetically-coupled centrifugals, the bearings and thrust washers wear. When the wear reaches a certain point, either the impeller starts to rub against the volute, or the inner magnet rubs against the containment can. Friction develops rapidly, and eventually it overburdens the magnets. At this point, the coupling breaks free; the outer magnet continues to turn, but the inner one stalls. Note that this is not a catastrophic failure; the fluid does not pour out of the pump. In fact, such pumps are often designed so that the impeller will contact the pump body before the inner magnet contacts the containment can — this prevents the can from being worn through.

When a pump using weak aluminum-nickel or ceramic magnets breaks free, the coupling gets demagnetized, and the entire pump has to be replaced. The better pumps normally used in CPI applications will not demagnetize. In these units, only the bearings and thrust washers, and perhaps the shaft, need to be replaced.

Canned-motor centrifugals have the same wear points, but they can fail catastrophically if the rotor and containment can come in contact: The shaft sheath and can are very thin, to keep the gap between rotor and stator to a minimum. If the rotor hits the can, it can tear through, damaging the motor windings and releasing fluid. Such damage cannot be repaired in the field.

The larger sealless gear pumps are designed so that the gears will contact the housing before the inner magnet contacts the can. This prevents catastrophic failure, and also saves on repairs — gears cost less to replace than magnetic couplings and containment cans.

In diaphragm pumps, performance falls off when the inlet and outlet valves wear down. Such pumps fail when one of the diaphragms bursts: If the process-side diaphragm gives out, hydraulic fluid gets into the process fluid. If the other diaphragm fails, the pistons or push rods can be damaged.

Pump economics

Sometimes, the application forces the user to select one specific type of pump — e.g., to avoid leaks, handle a certain viscosity, or produce a certain head. But in many cases the user can choose among two or three types of pumps, and

Table II — Sealless pumps generally cost less to maintain than sealed ones

	Time between failures, yr		
Seal	0.5	1	1
Bearing	1	1	0.5
	Annual cost of maintenance		
Sealless gear pumps			
Replace bearings	$80	$80	$160
Sealed gear pump			
Replace seal	$400	$200	$200
Replace bearings	80	80	160
Clean up spill	200	100	100
Product loss			
$0/lb	$0	$0	$0
$100/lb	200	100	100
Total	$680–880	$380–480	$460–560
Saving with sealless	$600–800	$300–400	$300–400
Sealless centrifugal pump			
Replace bearings	$60	$60	$120
Sealed centrifugal pump			
Total	$600–800	$300–400	$300–400
Saving with sealless	$540–740	$240–340	$180–280

here the choice should be made by evaluating their initial cost, power consumption, cost of maintenance, and probable cost of cleanup, repair and unplanned downtime due to failure. We will compare the various types of sealless pumps with each other, and with their sealed counterparts.

Table Ia shows approximate initial costs for plastic and stainless-steel sealless pumps; plastic centrifugals cost the least. If we were to compare the pumps based on cost per unit of capacity and pressure ($/gpm-psi), diaphragm pumps would be the least expensive. However, centrifugals are the most efficient.

For example: It takes 0.35 theoretical hp to pump a 20-cP liquid at 20 gpm with a 70-ft discharge head. Efficiencies for these circumstances are about 70% for a centrifugal pump, 45% for a gear pump, and only 20% for a diaphragm pump; thus the power requirements are 0.50, 0.78 and 1.75 hp for the three types. Table Ib shows what the difference in efficiency is worth over a five-year period: The centrifugal pump uses about $1,125 worth of power (8 h/d, 6¢/kWh), while the diaphragm pump uses $3,950.

As for sealless vs. sealed pumps, sealless ones generally cost more initially, except small centrifugals: Because the sealless ones are either pedestal-mounted or close-coupled, the user does not have to spend an extra $150–200 for a base or coupling. Table Ic compares the costs of centrifugals and gear pumps; the premium for sealless construction is greater for the centrifugals. The sealless pumps are about equal to the sealed ones in efficiency, but they cost less to maintain.

Maintenance and repair costs

Some pumps run for years before any parts have to be replaced; others have to be thrown away after a few days or even a few hours. But most pumps lie somewhere between these extremes: A pump that is designed properly, made of the right materials and correctly installed and lubricated, will typically run for six months to two years between maintenance shutdowns.

How do sealless pumps compare with sealed ones? Sealless centrifugal and gear pumps fail when the bearings fail. Repair involves tearing down the pump and replacing the bearings. For pumps under 20 gpm, new bearings cost about $10 apiece — centrifugals have two, and gear pumps have four. It takes an hour of labor ($40) to do the replacement. Sealed gear pumps have bearings as well as seals; centrifugal pumps have external bearings that do not typically fail.

In the event of a seal failure, there will be some leakage that has to be cleaned up ($100 labor). Also, some product will be lost — if a seal leaks 50 drops/min for 3 h, that adds up to 1 lb of product. A typical seal (for a ½-in. shaft) costs about $150, plus 15 min of extra labor above what it takes to replace bearings ($50 total).

Having these rough estimates in hand, let us look at the maintenance costs of sealed vs. sealless gear and centrifugal pumps. Table II sums up the results:

Gear pumps. When a sealed gear pump fails, both bearings and seals have to be replaced, and some product is lost. When a sealless gear pump fails, just the bearings have to be replaced, and there is no product loss. Assuming that bearings and seals fail at the same rate, sealed pumps cost about $300 more per failure than do sealless pumps, plus the value, if any, of the spilled product. For a once-a-year failure rate, this means an annual saving of $300 or more with the sealless pumps. This is enough to justify spending $900–1,500 extra for the sealless type (3–5-yr payback). The actual cost for sealless gear pumps under 20 gpm is $250–1,000 more than for sealed ones. If seals fail more often than bearings do (e.g., twice a year, vs. once), the payback for the sealless design is that much faster.

Centrifugal pumps. When a sealed centrifugal fails, only the seal needs to be replaced — the external bearings are easy to lubricate and thus should not wear out. If seals and bearings both fail once a year, the annual saving for a magnetically coupled design is $240 or more. For pumps in the 10–20-gpm range, the premium for a sealless design is about $600, on average. This translates to a 2½-yr payback for the sealless design.

Conclusion

Of course, the economics will vary from one application to another. If the product is particularly valuable, difficult to clean up, or hazardous, the sealless pump will save that much more. There is also the value of having a cleaner plant with fewer chemicals in the atmosphere. Here again, the sealless pump provides an advantage.

The author

Ali M. Nasr is president of Liquiflo Equipment Co., 140 Mt. Bethel Rd., Warren, NJ, 07060. The company manufactures corrosion-resistant pumps, both sealed and sealless, for chemical, petroleum and other process applications. Before joining Liquiflo, he worked in marketing and planning at Mobil Oil and Ciba-Geigy. He has a bachelor's degree in chemical engineering, and an M.B.A., from Rensselaer Polytechnic Institute.

SEALS FOR ABRASIVE SLURRIES

Liquids that contain abrasive particles in suspension, and concentrated solutions that may form abrasive crystals present special problems for the engineer who must select mechanical seals for pumps and the like. This article explores the various types of seals that are best for each of the several "families" of abrasive fluids that are encountered.

James S. Budrow, Durametallic Corp.

Products containing abrasive solids and those that tend to solidify or crystallize out are common to every major segment of the chemical process industries. Using mechanical seals for these services represents one of our most serious sealing challenges. Successful application of mechanical seals in such cases requires a clear understanding of the physical characteristics of abrasive fluids and their effect on seal performance.

The recommendations that are made in this article are based on both laboratory tests and over 50 years of field experience in sealing abrasive slurries.

Before we go any further, let us define a few terms:

Pusher-type seals — These seals use one or more springs to maintain contact between the seal faces. Pusher type seals have a dynamic gasket, usually an O-ring, to prevent leakage along the shaft. As the seal wears, this spring-loaded dynamic gasket moves forward slightly, to compensate for wear. Non-pusher-type seals doe not require a dynamic gasket, e.g., bellows-type seals.

Bellows-type seals — These seals are available with either metal or elestomer bellows. In either type, the seal ring is free to move forward (without contacting the shaft) as the insert wears. The elastomer bellows requires a concentric spring; the metal bellows does not because the bellows itself provides the pressure necessary to hold the seal faces in contact.

Why seals fail in abrasives

End-face mechanical seals operating in abrasive fluids are subject to five common modes of failure (Fig. 1). Understanding and controlling them is the key to a successful seal installation.

1. *Excessive seal-face wear*—Seal-face wear is accelerated by the grinding action of abrasive particles between the seal faces. Seal faces of carbon (graphite) versus ceramic will normally only last a few hours in highly abrasive fluids. To minimize wear, mating faces are usually chosen from materials with a hardness greater than that of the abrasives being sealed (seal-face materials of tungsten carbide or silicon carbide versus tungsten carbide). Excessive seal-face wear can be further combated by using: a clean external flush, double seals or a bypass flush through a cyclone separator or filter.

2. *Product side hangup*—This failure is evident when solids build up on the product side of the seal to the point that they impede or prevent free movement and tracking of the seal faces. Product side hangup immobilizes springs, pins and secondary seals. A nonclogging seal, such as a metal bellows type, and the use of large stuffing-box clearances help prevent product-side hangup.

3. *Atmospheric-side hangup*—Solids from normal seal-face leakage can accumulate on the atmospheric side of the seal. These solids can bridge the close clearances in many conventional seal designs, impeding free seal-face movement and tracking.

The use of non-pusher type seals with large shaft to I.D. clearances, and the presence of a buffer medium on the atmospheric side of the seal can help prevent this mode of failure.

4. *Erosion*—Erosion is the eating away, or washing out, of seal components in one localized area. It can result in excessive seal-face distortion or breakage of seal components. Erosion most frequently takes place on soft stainless steel or carbon materials. It can be visualized as a sandblasting effect on one localized area. Erosion can be dealt with by relocating the flush or reducing the

Originally published September 1, 1986

SEALS FOR ABRASIVE SLURRIES

flush rate in the stuffing box. Eliminating the abrasive particles in the flush medium (by the use of cyclone separators or filters) can be helpful in preventing such failure.

Another solution to this problem is the use of more erosion-resistant seal-face materials, such as tungsten carbide or silicon carbide.

5. *Drive-mechanism wear*—Wear on the drive-mechanism is accelerated, because the abrasive product acts as a grinding medium between the drive components as they rub together. Hardened drive components, moving the drive mechanism out of the product or eliminating the relative movement of the drive mechanism are ways of eliminating premature drive-mechanism failure.

Abrasive-fluid families

Abrasive fluids are categorized into families, based on the fluids' physical makeup or characteristics. Each family presents specific sealing problems. Hence, classifying an abrasive fluid into one of these families provides a general approach to sealing that application.

The abrasive-fluid families to be discussed include:

1. Nonfibrous abrasive slurries.
2. Fibrous slurries.
3. Dissolved solids.
4. Thermally sensitive fluids.

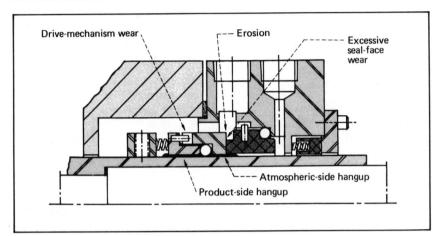

Figure 1 — Five common failure modes of seals in abrasive service

Nonfibrous abrasive slurries

Liquids carrying nonfibrous abrasive particles represent the most common type of abrasive application found in industry. Common slurries include drilling mud, diatomaceous earth, coal, phosphoric rock, clay and titanium dioxide. While each of these slurries is unique, they all present similar sealing problems. Typically, solids will pack up on the product side, or the atmospheric side, of the seal; also, seal-face and drive-mechanism wear is accelerated unless proper seal designs and auxiliary systems are applied.

Single metal bellows seals, as shown in Fig. 2, can typically be applied with good success in fluids with up to 10% solids (by weight). The smooth, uniform outer surfaces do not invite debris accumulation because the bellows rotation tends to sling the slurry out and away from the bellows convolutions.

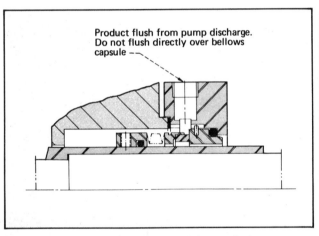

Figure 2 — Single metal bellows seal with bypass flush

Relatively large clearances on the bellows I.D. reduces the likelihood of the atmospheric-side hangup often encountered with conventional pusher-type seals. This can be further enhanced by filling the bellows I.D. with grease (Fig. 3) or providing a quench liquid to prevent the accumulation of these solids.

A bypass flush is recommended with single seals to prevent the slurry from packing up in the stuffing box. Care must be used with bellows seals to ensure that the flush is not located directly over the bellows, since the high velocity and direct impingement of the abrasive can result in rapid erosion of the thin bellows leaflets.

Slurry concentrations over 10% by weight may require external flushes and throat restriction devices or double seal arrangements. These arrangements provide an artificial environment for the seal by excluding the abrasive slurry from critical seal areas.

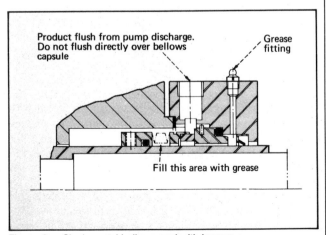

Figure 3 — Single metal bellows seal with bypass flush, and having the bellows interior filled with grease

Where product dilution can be tolerated, a single seal, throttle bushing in the bottom of the box, and clean external flush is highly effective (Fig. 4).

A sufficient flush rate must be provided, both to adequately cool the seal area and keep solids out of the stuffing box (Fig. 11).

Double seals are often required on high concentration slurry applications where product dilution is not permissible. Double seal arrangements (Fig. 5) use a clean buffer fluid between the seals, thereby reducing the likelihood of face wear and hangup. Double seals require that the buffer fluid pressure be maintained 15 to 25 psi above the product pressure at all times (Fig. 7). Where double seals are applied, it is good practice to provide hard faces on the inboard seal to prevent I.D. erosion.

If bellows seals are used, the abrasive slurry must be kept on the outside of the bellows. If the slurry is inside of the bellows, centrifugal force will force the solids into the bellows convolutions, plugging the bellows and causing possible failure.

Fibrous slurries

Many fibrous slurries have solids with specific gravities that are the same as, or slightly less than, that of the liquid medium.

As a result, the centrifugal force produced by the shaft and seal rotation does not always sling the fibers away from the seal. In fact, the opposite often occurs, which results in the stuffing box being packed with fiber.

Two common fibrous slurries are paper-stock and textile-dye liquors. All fibrous slurries require that there be a flush through the stuffing box to prevent fibers from packing up in the box.

Textile-dye liquors carrying synthetic or cotton fibers require only a single seal and a bypass fluid flush. Paper stock, on the other hand, readily plugs bypass flush lines and as a result, requires a clean external flush. In all cases, single nonclogging seals are preferred. Double back-to-back seals are not generally recommended because of potential fiber accumulation and plugging under the inboard seal. This varies with fiber density. Double seal arrangements which keep the product on the seal exterior are acceptable, although the stuffing box must still be flushed to prevent solids accumulation.

Dissolved solids

Fluids that contain dissolved solids are potentially abrasive—upon crystallization of the solids. This typically occurs when the product temperature is lowered or some of the fluid is evaporated off, so that the fluid medium can no longer retain all of the dissolved solids in solution, as with brine, plating solutions, caustics, carbonates, and sugar solutions.

These fluids present two major problems; seal-face abrasive wear and seal hangup. The simplest and best solution for preventing atmospheric-side hangup is either a single mechanical seal with a quench, a double seal, or a tandem mechanical seal.

In each case, a clean liquid is provided to wash away any product leakage, thereby preventing any accumulation (Fig. 6). Seal-face combinations of carbon-graphite and aluminum-oxide ceramic in pusher-type seals are generally acceptable as long as the problem of product crystallization is not present.

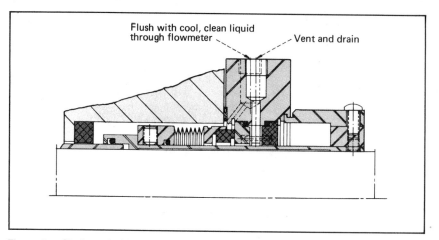

Figure 4 — Single seal with throttle bushing in bottom of box, with a clean, external flush

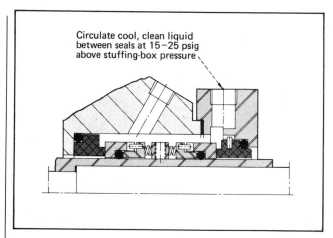

Figure 5 — Double seal, with clean buffer liquid between seals

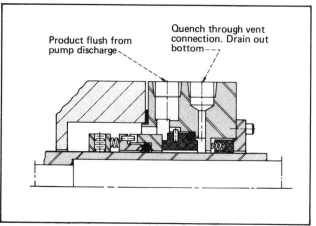

Figure 6 — Single seal with bypass flush, with quench through vent and drain.

Users who prefer single seals without a quench can employ pusher-type seals and face-material combinations, such as carbon versus aluminum-oxide ceramic in solutions containing up to 5% dissolved solids.

For from 5–10% dissolved solids, a single rotating bellows seal with hard faces, such as tungsten carbide versus silicon carbide is recommended. An intermediate quench in the vent and drain connections, to wash away solids, is helpful in extending seal life.

Fluids containing more than 10% dissolved solids may require special handling. This might call for a continuous quench, double, or tandem seals.

Thermally sensitive fluids

Thermally sensitive fluids are those that will solidify if temperatures are either raised or lowered beyond a given solidification point. They may or may not be abrasive in their liquid state. DMT (dimethyl terephthalate), black liquor, asphalt, and pitch all fall into this family of fluids. Some such fluids contain abrasive particles, in addition to being thermally sensitive.

Over the years, both single and double seals have performed well in these services. It is essential, however, that proper environmental controls be employed to prevent these liquids from solidifying. Failure to implement or

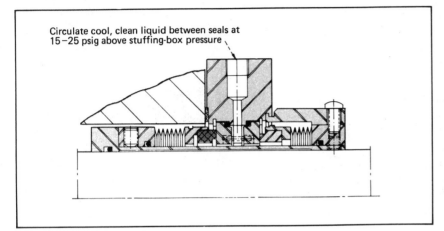

Figure 7 — Cartridge double seal with pressurized liquid between seals

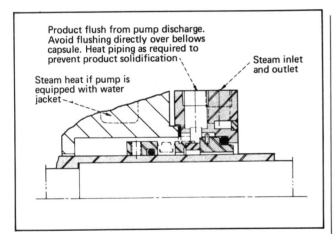

Figure 8 — Single seal with bypass flush and steam heating

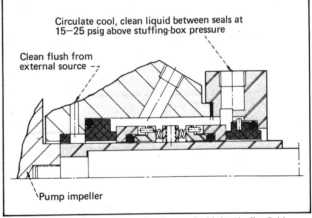

Figure 10 — Double seal can also be used with hot buffer-fluid and flush to avoid crystallization of heat-sensitive materials

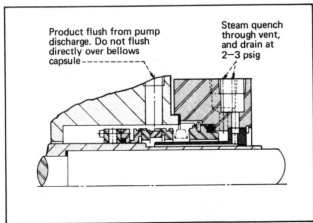

Figure 9 — Stationary metal bellows-type single seal with a steam-purge sleeve

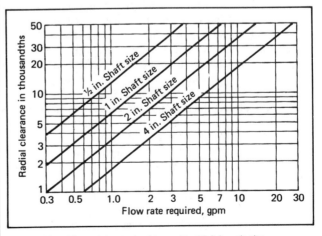

Figure 11 — Flowrates required to create 15 ft/s velocity

Table — Selection guide for seals intended to be used in various types and concentrations of slurries that contain abrasives

Product	Seal arrangement	Environmental controls	Fig.	Face materials
Non-fibrous slurry				
Less than 10% solids	Single rotating metal-bellows	Bypass flush	2	Tungsten carbide vs. silicon carbide
More than 10% solids	Single pusher type or single rotating metal bellows	External flush and throat bushing	4	Tungsten carbide vs. silicon carbide
	Double pusher type	Pressurized buffer fluid	5	Tungsten carbide vs. silicon carbide
Fibrous slurries				
Less than 10% solids and does not plug by-pass lines	Single rotating metal-bellows	Bypass flush	2	Tungsten carbide vs. silicon carbide
All others	Single metal bellows	External flush and throat bushing	4	Tungsten carbide vs. silicon carbide
Dissolved solids				
Less than 5%	Single pusher type or single rotating metal-bellows	Bypass flush	2	Carbon-graphite vs. ceramic
5–10%	Single rotating metal-bellows	Bypass flush	2	Tungsten carbide vs. silicon carbide
	Single pusher type or single rotating metal-bellows	Bypass flush and quench	6	Carbon-graphite vs. ceramic
More than 10%	Single pusher type or single rotating metal-bellows	Bypass flush and quench	6	Tungsten carbide vs. silicon carbide
	Double metal-bellows seal	Pressurized buffer fluid	7	Carbon-graphite vs. ceramic
Thermally sensitive				
	Single pusher type or rotating metal-bellows	Bypass flush, steam heated gland	8	Tungsten carbide vs. silicon carbide
	Single stationary high-temperature metal-bellows	Bypass flush, Steam purge bushing	9	Tungsten carbide vs. silicon carbide
	Double pusher type	Pressurized buffer fluid	5	Tungsten carbide vs. silicon carbide
	Double pusher type, or double metal-bellows seal	Pressurized buffer fluid and inner flush	10	Tungsten carbide vs. silicon carbide

Note: In pusher-type seals, sealing-face contact is maintained by one or more springs. In metal-bellows types, the bellows itself provides the pressure required to maintain face contact. On certain types of equipment, cartridge seals are recommended because they facilitate axial impeller adjustment.

maintain these controls frequently results in rapid failure, particularly with the single mechanical seal.

The key to applying single mechanical seals to thermally sensitive fluids is to assure that the product stays in the fluid state and does not reach a solidification point. Product solidification or thickening can result in excessive drive-mechanism wear, over-torquing of bellows convolutions, seal hang-up and seal face damage. To ensure that the product remains fluid, the stuffing box and gland areas are frequently heated with steam or heat transfer fluids (Fig. 8).

Where steam can be used in direct contact with the product, such as asphalt, a stationary bellows seal with a steam-purge sleeve is very effective (Fig. 9). The steam purge sleeve directs the steam under the bellows to the seal face area, effectively heating the box, cleaning the bellows inside surface, and preventing product oxidation (coking). Another advantage of using a stationary bellows is that the seal faces are placed deeper into the stuffing box, where the product is hot and less viscous.

Double seals, although requiring buffer-fluid systems, provide good seal-face lubrication and can help reduce start-up problems associated with product solidification around the seals.

In some very severe applications, such as hot coal slurries, care must be taken to prevent product solidification around the relatively cool inner seal. This can be accomplished by warming the buffer fluid or injecting a clean hot fluid on the product side of the inner seal (Fig. 10).

Proper startup and heatup procedures are a key to successful sealing of thermally sensitive fluids. Recommended time cycles vary depending on the size of the equipment, but in general, preheat of the equipment should start two or three hours before dynamic operation.

General recommendations for sealing various types of abrasive products are summarized in the above table. Inquiries from engineers who need more specific information on a given fluid and piece of rotating equipment should be directed to the seal-manufacturer's engineers.

The author

James S. Budrow is manager of application engineering with Durametallic Corp., 2104 Factory St., Kalamazoo, MI 49001; Tel: (616) 381-2650. Since joining the company in 1973, he has worked successively in R&D, design engineering and application engineering assignments. His work with abrasive slurries spans this period and led to key contributions in the application of metal bellows seals. He holds a B.S. in mechanical engineering from Western Michigan University, Kalamazoo.

Power consumption of double mechanical seals

Published pump efficiencies are generally based on the use of single mechanical seals during efficiency testing. If double seals are specified, the efficiency of the pump will be lower than that shown by the efficiency curves.

Frederic W. Buse, Ingersoll-Rand Corp.

The performance curves published by manufacturers of chemical process pumps are usually based on data obtained on pumps that have a single, multispring, mechanical seal in the stuffing box (Fig. 1). The curves are usually shown for 3,550 and 1,750 rpm. Many pump manufacturers run their tests at the design speed, 3,550 rpm, and use the affinity laws (summarized later) to calculate performance at 1,750 rpm.

In either case, there is seldom a correction factor or adjustment shown in the manufacturer's literature to account for the additional horsepower consumed by double mechanical seals. When a customer wants a double mechanical seal (Fig. 2), the quotes for efficiency are based on the published curve, not recognizing that the double seal will consume more horsepower than the single one.

As a result, the motor may be overloaded, especially when the pump is operating at 1,750 rpm. The pump efficiency can be one to ten percentage points below that quoted (Fig 3).

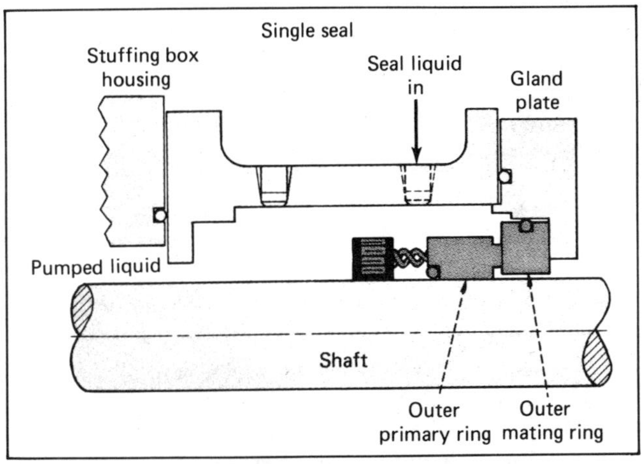

Testing seal power-consumption

Tests were run on pumps (without impellers) to determine the power consumption of the seals. These were unbalanced single seals, outside seals, and double seals, with carbon on ceramic mating surfaces (the most common materials), at pressures between 25 and 50 psi (the most common pressure range). A standard stuffing box was used, with flush water of 60° to 75°F injected into the seals for lubrication. Motor speed was varied from 1,100 to 3,600 rpm. Base horsepower was determined using a pump with no impeller, and no seal.

Note that the seal's power consumption will change with balanced (rather than unbalanced) seals, with a change in mating materials or mating-surface widths, or with an increase in stuffing-box pressure. Silicon carbide against carbon surfaces will have a higher coefficient of friction than carbon against ceramic, and tungsten carbide against carbon will have a still higher coefficient.

What the tests showed

There was no significant difference in power consumption between an unbalanced single inside seal and an outside seal. However, there was an increase in horsepower when double seals were used on pumps, especially those with low-horsepower drivers at four-pole speeds (Table I).

The additional power may seem insignificant, but when applied to pumps of low horsepower, there can be a substantial change in efficiency. The additional horsepower for the

Figure 1 — Single, multispring mechanical seal as used for pumps

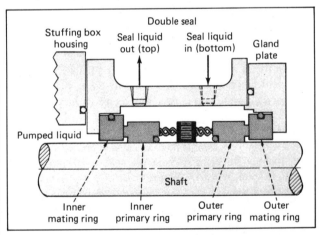

Figure 2 — Double mechanical seal consumes greater horsepower

Originally published December 9/23, 1985

double mechanical seal is added to the specified pump horsepower for the various sizes of pumps to obtain the new efficiencies. Table II shows the brake horsepower (bhp) range where there is a significant change:

To determine the change in pump efficiency for a given seal size, material and construction, *a constant difference in horsepower should be used*, not a constant decrease in points of efficiency (η). Fig. 4 shows pump horsepower at operating conditions vs. loss of efficiency when using double mechanical seals. (Fig. 4 is to be applied to pumps that have been tested at 1,750 or 3,550 rpm.)

Example 1

Find the loss for a double mechanical seal on a 3 × 1½ × 6 pump tested with a single 1⅛-in.-dia. seal, 1,750 rpm, 35 gal/min, 18-ft head, $\eta = 43\%$, liquid sp. gr. = 1.

$$bhp = QsH/3{,}960$$

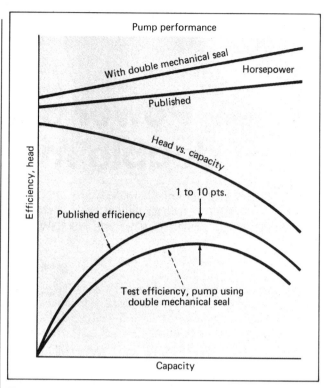

Figure 3 — Pump curves for single and double mechanical seals

Table I — How double seals increase power consumption

Speed, rpm	Additional horsepower consumption of unbalanced double seals of size (sleeve dia.), in.:	
	1⅛	1⅞
1,750	0.020	0.08
2,950	0.025	0.10
3,550	0.030	0.12

where Q = flow, gal/min; s = sp. gr., dimensionless; and H = head, ft.

bhp with single seal = $(35)(18)(1)/(3{,}960)(0.43) = 0.37$

From Fig. 4, the decrease in efficiency for 0.37 bhp is 2.3 points; efficiency with double seal = $43 - 2.3 = 40.7\%$.

bhp with double seal = $(35)(18)(1)/(3{,}960)(0.407) = 0.39$

Example 2

Find the loss for a double mechanical seal on a 3 × 1½ × 8 pump tested with a single 1⅞-in.-dia. seal, 1,750 rpm, 37 gal/min, 55-ft head, $\eta = 34\%$, sp. gr. = 1.

bhp with single seal = $(37)(55)(1)/(3{,}960)(0.34) = 1.5$

From Fig. 4, decrease in efficiency for 1.5 bhp is 3 points; efficiency with double seal = $34 - 3 = 31\%$.

bhp with double seal = $(37)(55)(1)/(3960)(0.31) = 1.66$
Horsepower increase = $(1.66 - 1.5)/1.5 = 10.7\%$

If totally enclosed motors were used, the 1.66-horsepower requirement would mean that a 2-horsepower motor would have to be employed, rather than a 1.5-horsepower one.

Effect of using affinity laws

The horsepower consumption of seals is even more significant if the data for a published curve is determined at one pump speed, but the performance for the speed in question has been predicted by use of the affinity laws. These laws state that capacity varies linearly with speed, that head varies with the square of speed, and power with the cube. That is, if you double the speed, you double the capacity, quadruple the head, and approximately octuple the power.

Likewise, when you reduce the speed from 3,550 to 1,750

Table II — Brake horsepower range where double seal is significant

Seal size (sleeve dia.), in.	Power below which there is substantial efficiency change, bhp	
	3,350 rpm	1,750 rpm
1⅛	2	2
1⅞	15	10

Table III — Additional bhp to be added when affinity laws were used

Seal size	Additional bhp for double seal*
1⅛	0.058
1⅞	0.287

*When stepping from 3,550 rpm performance to 1,750 rpm performance on the basis of the affinity laws.

Table IV — Decrease in efficiency when double seals are used

Typical bhp	Typical published η_p, %	Decrease, in points of efficiency for double seal of size shown:	
		1⅛ in.	1⅞ in.†
0.5	55	6.0*	
2	60		7.5
5	69		4.0
10	65		2.0

*Calculated in the example above.
†Calculated as in the example, but adding 0.287 bhp (from Table II) to the pump's bhp.

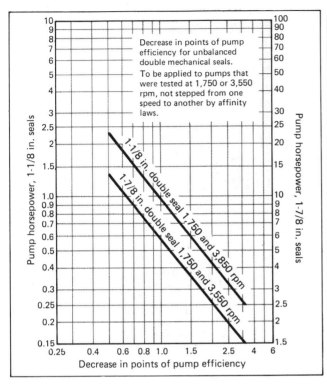

Figure 4 — Loss of efficiency for pumps tested at 1,750 or 3,550 rpm

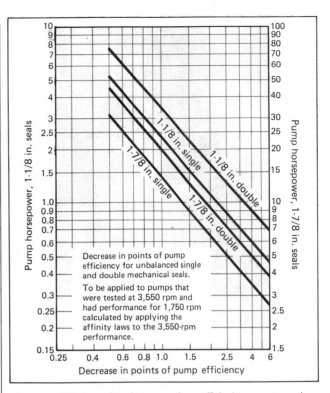

Figure 5 — Efficiency loss in cases where affinity laws were used

rpm, the capacity will be half, the head will be a quarter and the power an eighth of that when the pump operates at 3,550 rpm. These losses are from disk friction, discharges and mechanical friction losses. Traditionally, when operating at 1,750 rpm, two points were deducted from the efficiency at 3,550 rpm, to account for the extra losses. (When this rule was established, the losses for different types of mechanical-seal construction were not understood.) To give a more accurate value of the power at the lower speed, instead of taking one-eighth of the power at 3,550 rpm, one should calculate it by using the new capacity, new head, and efficiency at 3,550 rpm, less two points for additional losses.

Tests show that the horsepower consumption of a seal does not follow the affinity laws with change of speed; it is actually much closer to a linear relationship.

Table III shows the additional horsepower that would be added to the calculated pump horsepower that has been adjusted by two or three points for the change in speed.

Example 3

The following sample calculation shows how to determine the decrease in efficiency at 1,750-rpm performance for a double mechanical seal. The published efficiency, η_p, of 55% has already been adjusted 2 or 3 points for the lower speed.

Find the loss for a double mechanical seal on a 1½ × 1 × 6 pump, 1⅛-in.-dia. seal, bhp = 0.5, η_p = 55% efficiency.

From Table IV, for a 1⅛-in. seal, the loss for a double seal, $bhp_{seal} = 0.58$ bhp.

Efficiency with double seal = $(bhp_p)(\eta_p)/(bhp_p)(bhp_{seal})$

Efficiency with double seal = $(0.5)(55)/(0.5 + 0.58) = 49.3\%$

Decrease in efficiency, $\Delta\eta$, = 55 − 49.3 = 5.7 points.

Table III is a guide to the decrease in points of efficiency due to seal horspower consumption to be applied to performance that was calculated on the basis of the affinity laws, when stepping from 3,550 to 1,750 rpm.

Fig. 5 shows operating horsepower vs. additional decrease in points of pump efficiency when using a single or double seal, based on performance data that have been converted from 3,550 rpm to 1,750 rpm by use of the affinity laws.

Conclusions

There can be a significant increase in pump brake horsepower when using double mechanical seals on pumps driven by low-horsepower motors at low rpm. Depending on the type of seal and pump horsepower, the decrease in efficiency will vary from ½ to 10 points.

The increase in horsepower required should be applied to the operating pump horsepower to calculate the resulting efficiency. This differential is greater when pump horsepower has been stepped from one speed to another by using the affinity laws (rather than by actually measuring efficiency at the speed in question).

The author

Frederic W. Buse is chief engineer of the Standard Pump Div. of Ingersoll-Rand Pumps, P.O. Box 656, #1 Pump Place, Allentown, PA 18105, Tel: (215) 776-6100. He is a graduate of New York State Maritime College, with a B.S. in marine engineering. He is a member of the Hydraulic Institute (which named him "Man of the Year" in 1976) and of the American National Standards Institute (ANSI) committees B215 for centrifugal pumps, and B73.1 and B73.2 for chemical pumps. He has contributed to "Marks' Standard Handbook for Mechanical Engineers" and the "Pump Handbook" (ed. by Igor Karassik), and has just received his 11th U.S. patent on pumps.

Troubleshooting mechanical seals

A systematic method, based on failure analysis, for investigating and correcting the performance of mechanical seals provides the means for obtaining longer service lives and reduced costs.

William V. Adams, Durametallic Corp.

☐ Failure occurs when a product ceases to perform its intended purposes. Failure may occur after satisfactory life cycles have passed. Since equipment downtime is expensive and maintenance costs increase, whatever we can learn from failure analysis is frequently paid for many times over when corrective actions are taken.

In our discussions on failure analysis for mechanical seals, we will identify the:
- Basic components of a mechanical seal.
- Common causes for seal failures.
- Specialized observations and skills that can make seal-failure analysis a more exacting art.

Basic components of all seals

The function of every mechanical seal is to prevent the escape of a fluid past the clearance between a rotating shaft and the passageway through the wall of a housing or pressurized vessel. To do this, we will find, as shown in Fig. 1, that mechanical seals have three basic components: (1) a set of primary seal elements, (2) a set of secondary seals, and (3) hardware for attaching, positioning and maintaining face-to-face contact.

The primary seal is formed by two lapped faces that create a very difficult leakage path due to rubbing contact between them. In all such seals, one face is held stationary in a housing, and the other is fixed to and rotates with the shaft.

The phrase "very difficult leakage path" is used because all mechanical seals leak—even though one does not see leakage from most of them. The leakage rates, however, are normally small; and nonhazardous or nontoxic fluids may be allowed to evaporate or dissipate to the atmosphere in a short period of time. For hazardous and toxic fluids, other means are required for containment.

Leakage paths around the stationary face and the rotating face are usually closed by secondary seals of various fluoroelastomers. For pusher-type seals, the sec-

Originally published February 7, 1983

ondary seal must move forward along the shaft to compensate for wear and vibration at the seal faces. For nonpusher-type seals such as metal-bellows units, vibration and wear are taken up internally in the bellows, and here the secondary seals are truly static.

The mechanical hardware supplied with and integral to the seal is used to:

1. Adapt seals to various pieces of equipment. This hardware may consist of a sleeve or housing to make for an easier and more precise seal setting.

2. Provide mechanical preloads to the seal faces until hydraulic pressures can take over. This is normally accomplished by a large single-coil spring, or by a set of small coil springs.

3. Transmit torque to both stationary and rotating seal faces. This is normally accomplished by a series of drive pins, dents, notches or screws incorporated into the seal design.

No matter how complicated a design might appear, the first step in seal-failure analysis is to identify which of the basic seal components show damage that might indicate the cause of leakage.

Causes of failures

A mechanical seal has failed when leakage becomes excessive. The common causes of such failures include:

- Mishandling the components—Allowing the seal's components to become chipped, scratched or damaged prior to or during assembly.
- Incorrect seal assembly—This includes the incorrect setting, or misplacing, of the seal components in the seal cavity.
- Improper materials or seal design—This consists of selecting the wrong materials of construction or an incorrect design for the combination of pressures, temperatures, speeds and fluid properties required for a given application.
- Improper startup and operating procedures—This might be something as simple as failing to pressurize a double seal before starting a pump, or inadvertently running a seal dry.
- Fluid contamination—This might be the presence of harmful solid particles in the seal-cavity fluid.
- Poor equipment conditions—Here the problem might be caused by excessive shaft runout, deflection or vibration.
- Worn-out seal—The seal may have completed a satisfactory life cycle.

Learning via failure analyses

The objective of failure analysis, naturally, is to learn from failures. We should carefully look at worn and damaged seal parts, the condition of the equipment, and the operating conditions, in order to establish a list of ways to improve seal life.

For worn parts, this consists of identifying the damage as chemical, mechanical or thermal, and taking appropriate steps to ensure that the damage does not recur. Skills in mechanical-seal failure analysis can be improved by looking at the basic forms of chemical, mechanical and thermal damage that occur to mechanical seals, and determining:

1. What the damage looks like.
2. How the damage affects seal performance.
3. What the types of damage indicate about a seal's past history.
4. What corrective steps can be taken to eliminate various types of damage from occurring again under the same operating conditions.

Let us begin our analysis by discussing the symptoms, examining the causes and reviewing the corrective actions to be taken for failures of mechanical seals by chemical action.

Overall chemical attack

Symptoms—Seal failure from this cause will leave the parts appearing dull, honeycombed, flaky, or starting to crumble or break up, as shown in Fig. 2. When weight and material-hardness readings are taken on the damaged parts and compared with those on the original parts, they will prove to be substantially reduced.

Causes—This type of failure is due to corrosion, and is caused by using the wrong materials of construction for the chemical environment. If double seals have been

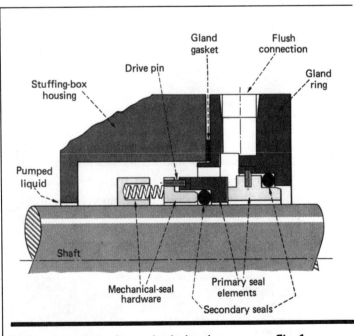

Basic components of a mechanical seal Fig. 1

132 MECHANICAL SEALS

Chemical attack produces overall corrosion of seal components — Fig. 2

used, it is a signal to check the reliability of the pressurizing system or the purity of the buffer fluid.

Corrective actions—

1. Obtain a complete chemical analysis of the product being sealed, and upgrade the materials of construction in the seal for the environment.

2. Neutralize the corrosive environment by using double seals; or, when using a single seal having a bushing or a lip seal in the bottom of the seal cavity, flush the seal from an external source with a clean, compatible fluid.

Fretting corrosion

*Symptoms—*Fretting corrosion is probably one of the most common types of corrosion found in mechanical seals. It causes leakage past secondary seals and leaves the shaft or sleeve that is directly beneath the secondary seal corroded and damaged. This area will appear pitted or bright and shiny in comparison with the overall finish of the rest of the shaft or sleeve (see Fig. 3).

*Causes—*Motion between two surfaces normally fixed with respect to one another causes fretting corrosion. In mechanical seals, fretting is due to a constant back-and-forth movement of secondary seals over a shaft sleeve—removing the passive coatings that would normally protect the sleeve. Constant vibration of the shaft packing over this surface also removes the passive coating and allows further corrosion to occur.

*Corrective actions—*We should evaluate the following options in order to reduce or eliminate damage occurring from fretting corrosion:

1. Check for excessive vibration in secondary seals. This can be done by making sure that shaft runout, shaft deflection and axial end-play on the equipment are held to a maximum of 0.003 in. (0.076 mm) total indicated runout.

2. Apply protective coatings of hard-facing alloys, chrome oxide or aluminum oxide directly under the area where secondary seals slide.

3. Upgrade the base material of the shaft or sleeve to a material that does not depend upon passive or protective coatings for corrosion resistance—for example, by using titanium.

4. Eliminate the use of Teflon V-rings, wedge rings and taper rings in favor of elastomer O-ring secondary seals. O-ring secondary seals are less susceptible to fretting corrosion because they are softer and are able to take up minor axial-shaft movement internally in the elastomeric material.

5. Switch to a nonpusher-type seal such as a rubber-, Teflon- or metal-bellows seal where the secondary seals are truly static.

Chemical attack on O-rings

*Symptoms—*Chemical attack can be suspected if the O-rings are swollen or have taken a permanent set that

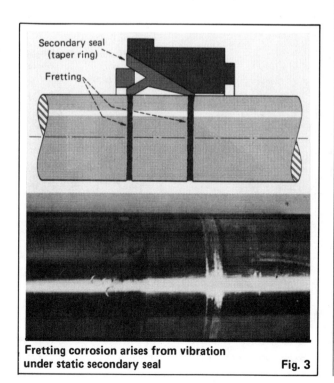

Fretting corrosion arises from vibration under static secondary seal — Fig. 3

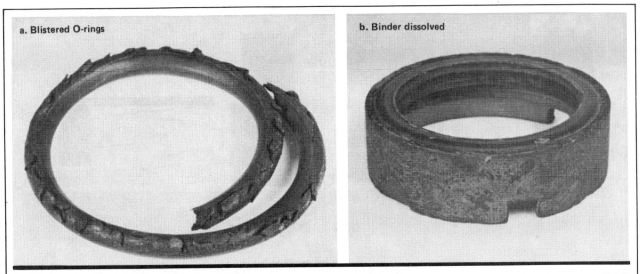

Chemical attack on the materials of construction for O-rings and sealing rings — Fig. 4

prevents axial movement of the sliding seal face. Chemical attack can leave O-rings hardened, bubbled, blistered on the surface, or with the appearance of being eaten away or breaking up (Fig. 4a).

Causes—Incorrect material selection, or loss or contamination of the seal buffer fluid.

Corrective actions—Make a complete chemical analysis of the liquid being sealed, and reevaluate the O-ring selection. These should be the first steps in analyzing this failure. Frequently, the presence of trace elements, originally overlooked when specifying the seals, will be at fault. If a suitable material cannot be found, create an artificial environment by flushing the seal from an external source.

Leaching

Symptoms—Leaching normally causes a minor increase in seal leakage and a sharp increase in wear of the carbon faces. Ceramic and tungsten carbide faces that have been leached will appear dull and matted (Fig. 4b)—even though no coating is present on them. Hardness readings on such seal faces will show a decrease of 5 points, or more, on the Rockwell A scale, from the original values.

Causes—Leaching occurs by chemical attack on the binder that holds the base material together in powdered-metal and ceramic materials. This type of attack can occur from a few ten-thousandths to several thousandths of an inch deep, and leave the seal parts beyond repair.

For example, caustic and hydrofluoric acid solutions will leach out 5%, or more, of the free-silica binders in ceramic seal rings. This will result in an excessive rate of wear of the carbon-seal face. If this wear is allowed to continue, aluminum oxide particles will actually be dislodged from the ceramic-seal face, cause further abrasive damage between the seal faces, and result in very short seal life.

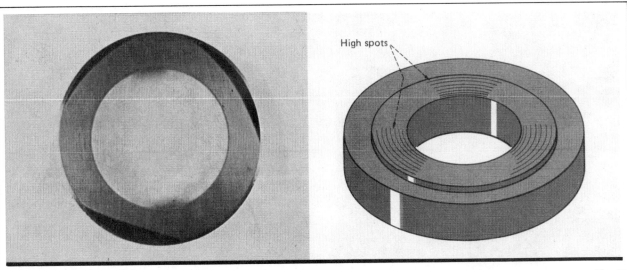

Distortion of the seal faces creates uneven wear, causing leakage — Fig. 5

134 MECHANICAL SEALS

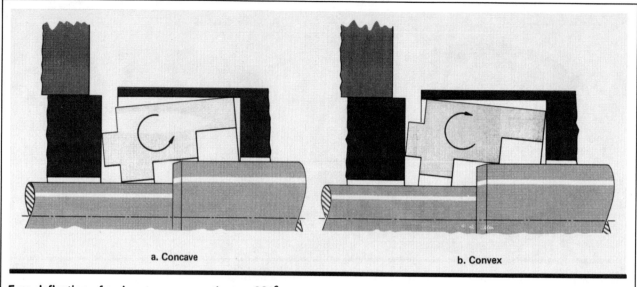

a. Concave   b. Convex

Face deflection of seals occurs as a continuous 360° pattern   Fig. 6

Corrective actions—Two procedures can be followed:

1. Upgrade the base material of the seal by using a higher-purity aluminum oxide such as 99.5% aluminum oxide for applications in caustic or hydrofluoric acid solutions. For cobalt-bound tungsten carbide materials that leach in the presence of water and other mild chemicals, change the binder from cobalt to nickel to eliminate chemical attack.

2. Use a seal arrangement that will provide a buffer fluid at the seal faces—e.g., a single seal with a flush stream from an external source, or a double seal with a suitable buffer-fluid system.

We will now consider seal failures that arise from mechanical problems.

Face distortion

Symptoms—Excessive leakage at the seal. Visual examination of the seal faces shows a nonuniform wear pattern. Sometimes this condition is difficult to detect. By lightly polishing the seal faces on a lapping plate, high spots will appear at two or more points, which indicates an uneven wear pattern (Fig. 5).

Causes—A number of factors are responsible for seal-face distortion:

1. Improper assembly of seal parts, causing nonuniform loads at two or more points around the seal face. This frequently occurs with rigidly mounted or clamp-style seal faces because uneven torquing of the gland nuts may transmit uneven deflections directly to the seal faces.

2. Improper cooling, inducing thermal stresses and distortions at the seal faces.

3. Improper finishing and processing of the seal parts, leaving them with a saddle surface or with high spots at several points around the seal faces.

4. Improper gland support, resulting from debris or

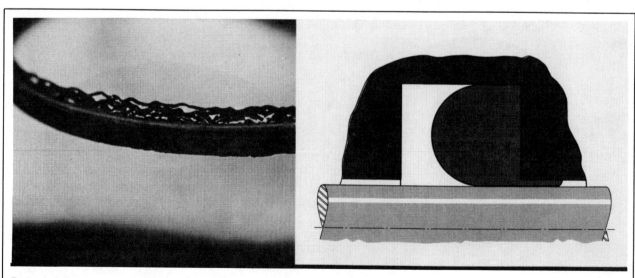

Extruded O-rings appear cut or peeled due to squeezing past very-small clearances   Fig. 7

deposits left in the gland, and/or physical damage that upsets metal in the gland and transmits an uneven load to the stationary seal face.

5. Poor surface finish at the face of the stuffing box, due to corrosion or mechanical damage.

Corrective actions—

1. Relap the seal faces to remove the cause of distortion.

2. Consider the use of flexibly mounted stationary-seal faces to compensate for gland distortion.

3. Readjust the gland by positioning the gland nuts finger-tight, then torquing them evenly to an appropriate value.

Seal-face deflection

Symptoms—Uneven wear at the seal face, like that for face distortion. However, the wear pattern is continuous for 360 deg around the seal faces, and is concave or convex. A convexed seal face will result in abnormally high leakage rates, while a concaved one may result in excessive seal-face torque and heat (Fig. 6). Seals exhibiting either condition will generally not be stable under cyclic pressure conditions.

Causes—Seal face deflection may arise from:

1. Improper stationary-seal-face support.

2. Swelling of secondary seals.

3. Excessive deflection of seals when they operate beyond their pressure limits.

4. Inadequate balancing of hydraulic and mechanical loads on the primary seal faces.

Corrective actions—

1. Check the operating limits for the seal design.

2. Consider flexibly mounting the stationary seal.

3. Incorporate seal-face materials (such as bronze, silicon carbide or tungsten carbide in place of carbon) having a higher modulus of elasticity. These will have greater resistance to hydraulic and mechanical bending loads.

Extrusion

Symptoms—O-rings or other secondary seals show deformation from being squeezed past the close clearances around the primary seal faces. Frequently, these O-rings or secondary seals will appear cut or, in some cases, peeled (see Fig. 7).

Causes—Excessive temperatures, pressures or chemical attack, making the O-ring soft, or excessive stresses in the O-ring for the given clearance.

Corrective actions—

1. Check O-ring clearances for the application (see data in Fig. 8).

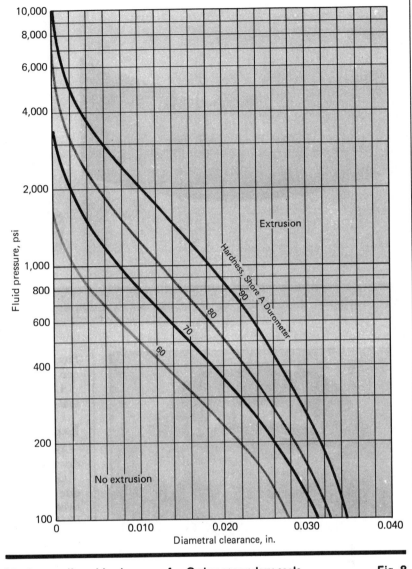

Maximum allowable clearance for O-ring secondary seals   Fig. 8

2. Check the chemical compatibility and temperature limits of the secondary seals.

3. Install anti-extrusion rings if necessary.

Erosion

Symptoms—Seal face eaten away or washed out in one localized area (Fig. 9). Erosion will commonly occur with the stationary seal face until excessive seal-face distortion or breakage occurs. Erosion most often takes place in carbon-graphite material but can arise in other materials under more severe conditions.

Causes—Excessive flush rates, or normal seal-flush rates with a flush fluid contaminated with abrasive particles. Both conditions will result in a sandblast effect on a localized area of the stationary seal face.

Corrective actions—

1. Reduce the seal-flush rate.

2. Eliminate the presence of abrasives in the seal-flush fluid by using filters or cyclone separators.

3. Switch to more-erosion-resistant stationary seal-

136 MECHANICAL SEALS

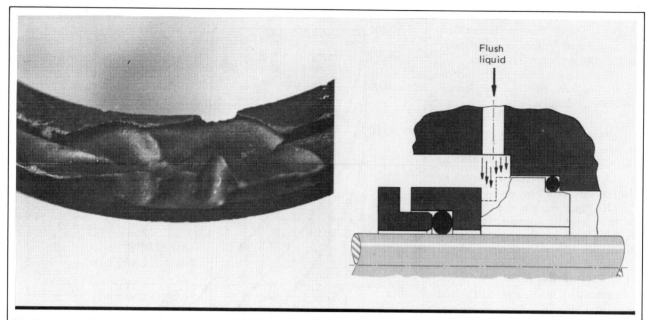

Erosion caused by excessive flushing or abrasives usually dissolves the stationary seal face — Fig. 9

Excessive wear on drive pins, drive dents and drive slots — Fig. 10

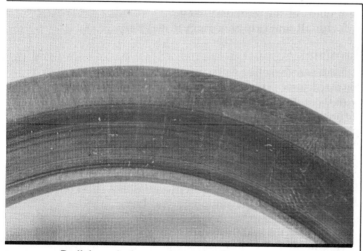

Radial cracks in metal or ceramic rings indicate heat checking — Fig. 11

face materials such as bronze, tungsten carbide or silicon carbide in place of carbon.

4. Relocate the seal flush, or shroud the stationary seal face from the direct flow of the flush.

Excessive drive-pin wear

Symptoms—Excessive wear of drive pins, drive dents or drive slots in a short period of time (Fig. 10).

Causes—Rapid wear can occur on drive mechanisms, due to heavy loads and large degrees of movement between the drive mechanism and other wear surfaces. High wear rates may also occur with relatively little movement if the drive mechanism is not properly lubricated. For example, drive mechanisms that operate in dry-nitrogen or dry-air environments containing abrasive particles will wear more quickly than those in a clean environment, or those operating with an oil or water lubricant. Similar conditions arise with drive mechanisms in liquids contaminated with abrasives vs. clean liquids. The major cause of heavy drive-mechanism wear is excessive face runout at the shaft-to-stuffing-box location.

Corrective actions—

1. Check the condition of the equipment and limit shaft end-play, shaft deflection and out-of-squareness of the shaft with respect to the stuffing box to 0.003 in. (0.076 mm) total indicated runout.

2. Incorporate hardened drive pins or drive dents in the seal design.

3. Consider seal designs that will put the drive mechanism in a better lubricating environment—e.g., substituting double seals for single seals.

4. Check pressure limitations of the seal design.

Let us now discuss thermal failures.

Heat checking

Symptoms—The presence of fine to large radial cracks that seem to emanate from the center of the metal or

Vaporization shortens seal life and impairs seal performance   Fig. 12

ceramic ring (Fig. 11). These cracks act as a series of cutting edges against carbon-graphite and other seal-face materials, and this scraping action very rapidly wears out the seal.

Causes—The common causes for heat checking include: (1) lack of proper lubrication, (2) vaporization at the seal faces, (3) lack of proper cooling, and (4) excessive pressures and velocities. Any one or a combination of these factors can result in higher friction and heat at the seal faces. The excessive thermal stresses thus developed will result in fine fractures.

Corrective actions—

1. Check operating conditions for the application, and make sure that these are within the prescribed limits for the seal.

2. Confirm that adequate cooling and flow are available at the seal faces to carry away the seal-generated heat. Rule-of-thumb guidelines are that (a) the temperature of the fluid flowing through the seal cavity should not exceed a 40°F (22°C) temperature rise, and (b) pressure of the seal-cavity product should be maintained 25 psi (1.72 bar) above the vapor pressure of the seal-cavity fluid to avoid vaporization.

3. Check to make sure that the seal has not been overloaded. The problem here is whether a thrust bearing or thrust collar in the equipment may have become damaged or inoperative, thereby creating excessive seal-face loads.

4. Upgrade the seal-face materials. For example, if hard-facing alloys are used, substitute tungsten carbide or silicon carbide, having higher pressure-velocity (*P-V*) limits and resistance to heat checking.

5. Reduce the *P-V* value of the seal. This is a factor of the pressure (psi) at the seal faces, multiplied by the velocity (ft/min) of the outside diameter of the seal face. It is possible to consult the manufacturer and obtain a revised dimension that will reduce the hydraulic load at the seal faces to provide a lower *P-V* for the same face materials.

6. Check the cooling and lubrication at the seal faces. If necessary, improve them.

Vaporization

Symptoms—Popping, puffing and blowing of vapors at the seal faces is known as vaporization, and results in excessive leakage and damage to them. If vaporization does not cause catastrophic failure, it usually shortens seal life and impairs seal performance. Inspection of the seal faces frequently shows signs of chipping at the inside and outside diameters, and pitting over the entire area (Fig. 12).

Causes—Vaporization occurs when heat generated at the seal faces cannot be adequately removed, and the liquid between them flashes. Vaporization can also be caused by operating the seal too near the flash temperature and flash pressure of the product in the seal cavity. Other operating conditions that will bring about vaporization include:

1. Excessive pressure for a given seal.
2. Excessive seal-face deflection.
3. Inadequate cooling and lubrication of the seal.

Vaporization may be an indication that a seal flush has become inoperative, or that cooling water to a heat exchanger has been shut off or reduced.

Corrective actions—

1. Improve circulation and cooling at seal faces.
2. Make sure the seal is operating at temperatures and pressures well below the flash conditions of the product in the seal cavity.
3. Check the seal design to see if it is being used within its pressure and speed limits.
4. Consult the seal manufacturer for recommendations on reducing self-generated heat.

Rule-of-thumb limits indicate that the operating temperature and pressure at the seal should be at least 25°F (14°C) and 25 psi (1.72 bar) lower than the flash temperature and flash pressure of the product in the seal cavity.

Blistering

Symptoms—Blistering (Fig. 13) is indicated by small circular sections that appear raised on the carbon seal faces. Sometimes, this condition is best observed by

138 MECHANICAL SEALS

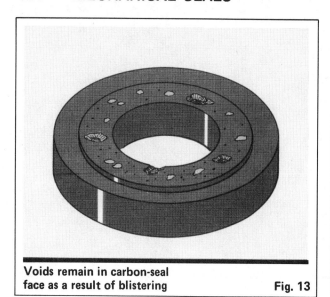

Voids remain in carbon-seal face as a result of blistering Fig. 13

using an optical flat, or lightly lapping the seal faces. These blisters separate the seal faces during operation and cause high leakage rates. Blistering usually occurs in three stages:

Stage I: Small raised sections will appear at the seal faces.

Stage II: Cracks will appear in the raised sections, usually in a starburst pattern.

Stage III: Blisters will be pulled out, leaving voids in the seal face.

Causes—The exact cause of blistering is still somewhat debatable. The best explanation is that high-viscosity fluids such as SAE #10 oils penetrate the interstices of the carbon seals over an extended period. When the seals heat up, the oil is suddenly forced out of the pores. Blistering commonly occurs in seals that start and stop frequently in high-viscosity fluids.

Corrective actions—

1. Reduce the viscosity of the fluid in the seal cavity either by substituting a new seal-cavity fluid or by increasing the fluid's temperature.

2. Eliminate frequent starts and stops of equipment containing mechanical seals.

3. Substitute a nonporous seal-face material such as tungsten carbide, silicon carbide or bronze for the carbon-graphite.

4. Check cooling and circulation to the seal faces. Improper cooling and circulation will makes seals more susceptible to blistering.

Spalling

Symptoms—Similar to blistering but occurring on surfaces away from the seal face, such as the outside diameter and the back side of the seal (Fig. 14).

Causes—Spalling, like blistering, is caused by excessive thermal stresses in a carbon-graphite seal. Unlike blistering, however, spalling seems to occur with virtually any fluid, and is the result of moisture being suddenly driven off when the seal is overheated. Spalling is almost exclusively due to dry running of the seal. So if seal parts are heavily spalled, it is a good indication that the equipment was allowed to run dry for an extended period of time.

Corrective actions—To avoid dry running of a mechanical seal, a pressure or load switch should be added to the equipment. Or, an alternative sealing method should be supplied, such as a double seal having a thermal-convection or forced-lubrication system.

O-ring overheating

Symptoms—When overheated, elastomer O-rings harden, crack and become very brittle. Secondary seals of Teflon will harden, tend to discolor (becoming bluish-black or brown), show signs of cold flowing, or take the shape of the secondary-seal cavity.

Causes—Overheating is generally due to lack of cooling or of adequate flow in the seal cavity. It can also be caused by excessive temperatures, or simply by an incorrect selection of materials.

Corrective actions—If overheating of O-rings is noted:

1. Check cooling and flow in the seal cavity area—including the lines for blockage, and heat exchangers for buildup of scale.

Spalling is similar to blistering but occurs on surfaces other than seal face Fig. 14

Varnish or abrasive sludge settles on atmospheric side of mechanical seal   Fig. 15

2. Apply cooling. If the temperatures are still excessive for a given elastomer secondary seal, consider a metal-bellows seal having higher temperature limits.

Oxidation and coking

Symptoms—Oxidation and coking leave a varnish, lacquer or abrasive sludge on the atmospheric side of the seal (see Fig. 15). This can cause rapid wear of the seal faces and/or hangup of both pusher and nonpusher types of mechanical seals.

Causes—Coking is the result of oxidation or chemical breakdown of hydrocarbons to form heavy residues.

Corrective actions—

1. Apply a steam purge to the atmospheric side of pusher and nonpusher mechanical seals to carry away sludge or abrasive debris.

2. Flush the seal from a clean, cool external source in order to eliminate coking in the seal cavity.

3. Apply cooling to the seal cavity by using a stuffing-box water jacket, or a water-cooled or air-cooled heat exchanger.

4. Switch from carbon to hard seal-face materials that withstand the abrasive action from particles formed by oxidation, and purge the seal on the atmospheric side with steam to remove sludge and debris.

In general, hydrocarbons should be cooled to below 250°F (121°C) in the seal cavity to prevent coking and oxidation. This temperature limit depends greatly on the fluid handled. For example, the oxidation limits for heat-transfer fluids are above 350°F (177°C).

In summation

Failure analysis is not always straightforward and exact, but it does follow a systematic approach:

Step 1—Identify the problems resulting in short seal life. This may not always be a seal-design problem.

Step 2—Carefully examine possible solutions to the problem. Past experience, feedback from equipment manufacturers, and consultation with a knowledgeable seal expert will be valuable in formulating a list of possible answers.

Step 3—Choose an appropriate remedy and take corrective action. Selecting the best one will require an analysis of cost, availability of hardware, and future economic benefits.

Step 4—Follow up on the problem-solving efforts.

References

1. "Metals Handbook," 8th ed., Vol. 10, Failure Analysis and Prevention, American Soc. for Metals, Metals Park, Ohio, 1975.
2. "Guide to Modern Mechanical Sealing," 7th ed., Durametallic Corp., Kalamazoo, Mich., 1979.
3. Catalog ORD-5700 [O-rings], Parker Hannifin Corp. Seal Group, Lexington, Ky., 1977.
4. Strugala, E. W., The Nature and Cause of Carbon Blistering, *ASLE Trans.*, Vol. 28, pp. 333–339, American Soc. of Lubrication Engineers, Park Ridge, Ill., 1972.
5. "Process Industries Corrosion," p. 24, National Assn. of Corrosion Engineers, Houston, 1975.

The author

William V. Adams is Director of Engineering for Durametallic Corp., 2104 Factory St., Kalamazoo, MI 49001, where he is responsible for the supervision of design, application and drafting personnel. Mr. Adams is past chairman of seal-education programs for the American Soc. of Lubrication Engineers, U.S. Dept. of Energy, and ASME. He has a B.S. in mechanical engineering from Western Michigan University, and is a member of the American Soc. of Lubrication Engineers and American Soc. for Testing and Materials.

Part V
Pump drives

New turn for CPI motors
Specifying electric motors
Check pump performance from motor data
Making the proper choice of adjustable-speed drives

New turn for CPI motors

Sophisticated electronic control devices, a new magnet material, and a proposed design standard for chemical-duty motors may succeed in raising overall motor performance.

☐ Ever-higher efficiency is the big trend in the world of medium-horsepower electric motors. And the chemical process industries (CPI) are reaping the benefits. Since the late 1970s, CPI firms have been switching to energy-efficient (EE) motors (*Chem. Eng.*, Nov. 17, 1980, p. 93). More recently, industry experts have been trying to write a motor-design standard exclusively for CPI plant applications. And now they are testing the first of a new wave of devices that, relying mostly on sophisticated electronics, promise improved motor performance and smoother operation.

Most of the new efficiency-boosters work by closely controlling either motor speed or power factor (ratio of power used to power delivered to motor). But it is still too early to tell whether they will live up to manufacturers' claims. And it may take some doing to persuade the CPI that they should try some of the devices—especially new power-factor controllers. "Our company has tested a number of these units through the years, and we have yet to find one that does what its manufacturer says it does," observes one CPI motor expert.

Other engineers who know about motors are more optimistic about the new electronic offerings. "Electronics is very competitive and rapidly changing, and there have been many technical breakthroughs that reduce costs. It really looks like microchips are the way to go," says Rod Van Horn, a senior electrical engineer with Dow Chemical Co. (Midland, Mich.).

The mood seems to have caught on with manufacturers, who last year announced a number of developments. Here is a partial list:

■ General Electric Co. (Schenectady, N.Y.) unveiled a "programmable motor" that combines the variable-speed ability of d.c. motors with the reliability and durability of a.c. units. A similar design by Tasc Drives Ltd. (Lowestoft, U.K.) is about to be marketed in the U.S.

■ Harris Corp., a Melbourne, Fla., semiconductor manufacturer, began marketing a customized microprocessor chip that functions as a motor controller (for single-phase motors). Notable among its innovations is the ability to handle up to 3,500 V on the power line—a very high voltage for microchips.

■ Chesebrough-Pond's, Inc. (Greenwich, Conn.) developed a relatively inexpensive microprocessor-based motor controller that incorporates many of the attributes of a power-factor controller devised at the National Aeronautics and Space Administration (NASA) several years ago.

■ General Motors Technical Center (Warren, Mich.), Sumitomo Special Metals Co. (Osaka) and others are in a race to market a new type of magnet material—composed of neodymium, boron and iron—that may enable motormakers to convert many present wound-stator designs into more-energy-efficient permanent-magnet versions.

THRIFTY GOAL—These and other developments aim at cutting the energy spent in running motors. A pair of studies done during the late 1970s for the U.S. Dept. of Energy found that 58% of all electrical energy generated in the U.S. goes into powering motors, and that over 45% of that energy is used in manufacturing alone; this represents a large potential for conservation.

More to the point, DOE found that the chemical industry, on the average,

New magnet material is made by rapid quenching of liquid mixtures

Originally published April 2, 1984

spends 2.2% of its revenues on motor power, while manufacturing as a whole spends only 1.1%. Moreover, the DOE Energy Information Administration's latest study (1983) indicates that the cost of electricity will continue rising in constant 1982 dollars over this decade—from about 5.66¢/kWh in 1984 to 6.07¢ in 1990.

The careful pondering of such figures may have played a role in the switch to so-called EE motors, which, though more expensive than conventional ones, feature better designs and materials of construction. Also credited with spurring the production of such units is a motor-efficiency guideline developed by the National Electrical Manufacturers Assn. (NEMA) in 1981; it standardizes the testing procedures that measure efficiency.

A 1982 market study by Predicasts, Inc. (Cleveland) indicates that a price premium of 25% is typical for most EE motor sizes. But payback analyses show that the price differential is relatively quickly returned in lower costs.

"Sales of EE motors have taken off, and we're quite pleased with how things have turned out," says a NEMA spokesman. Jeff Cosman, a GE sales manager, says NEMA market studies show that one out of six motors sold in the U.S. in 1983 had an EE design, up from one in eight in 1982. "I would be surprised if EE motors accounted for less than half of those bought by the CPI, because of the greater potential for savings there," he adds.

A motor expert at a major chemical company feels the proportion should be much higher. "I recommend installing EE motors everywhere in the company," he says. "They may not always be economically justifiable, but for inventory purposes this is the better way to go."

MICROCHIP CONTROL—While the energy savings from EE motors are substantial, they are often surpassed by those obtained from motor-speed controllers, especially variable-frequency drives (see the article on p. 161). Varying the speed by

Chesebrough-Pond's motor controller also functions as soft starter

electronically controlling power input can reduce costs substantially, particularly in applications such as fans, blowers, and pumps of varying flows. But the drives traditionally have had limited appeal in the CPI because they are expensive, with costs ranging into hundreds of dollars per unit of motor horsepower.

One type of motor control that has potential for saving energy is the Nola-type controller,* based on a design invented by Frank Nola, a NASA scientist.

The original Nola controllers (early commercial versions have been available since around 1980) have had very little success in the CPI, mostly because the control device has not been sensitive enough to gauge motor loads. But Harris Corp., for one, has succeeded in integrating the sensing and control capabilities into a microchip called the HV-1000. As yet, the device is available only for single-phase motors (not the polyphase ones commonly used in the CPI). Harris says the chip is the first in the world that can run on regular 110-V or 230-V lines without the need for electrical buffers to moderate the current.

According to Harris, the ideal applications are those in which motors of more than ¼ hp are run with vary-

*The device has also been called a "power factor controller," which is something of a misnomer because it doesn't actually control the power factor, but senses when it is very low and interrupts the current momentarily. Because a low power factor corresponds to a lightly loaded motor, the current-cutting saves some energy when the motor is not running at full speed.

ing loads. The microchip senses the power factor, and a triac (an electronic switching device) cuts current until the voltage and current are nearly back in phase (yielding essentially a power factor of 1). Power savings are 10% or more vs. conventional motors, according to the firm.

ADDED VERSATILITY—The same concept, but in a much different design, is incorporated into a device developed at Chesebrough-Pond's, though for three-phase designs as high as 75 hp. It includes an Intel Corp. 8048 microprocessor that has an algorithm built into it to calculate power factor.

"Energy saving across the board is between 2% and 5%, but it is not the only thing this device is capable of," says Michael Westkamper, manager of the firm's advanced technology group. He adds that the device can also serve as a "soft starter," in cases where the motor must be brought up to speed gradually—such as when a loaded blender is started cold. Having soft-start capability is a particular advantage in extrusion equipment, where high pressures are developed as the motor slams the extrusion piston. "Our maintenance engineers have found that seal life is vastly extended," says Westkamper.

At Chesebrough-Pond's, about 25 of the devices already have been installed on motors powering air chillers, fans, blenders and other devices, and about 150 more will be installed this year. Westkamper notes that there are other soft-starting methods available, such as part-winding-start motors† or electronic soft starters. But he asserts that the former are inherently inefficient (because of the greater electrical resistance of the partial winding, as opposed to that of a conventional rotor), and that the latter can cost several hundred dollars/hp, vs. the less-than-$100/hp expected for production quantities of the Chesebrough-Pond's design.

The company had licensed its design to National Semiconductor Corp. (Santa Clara, Calif.) to make a

†A motor that starts with only half of its winding energized.

single-phase version on a microchip—i.e., similar to what Harris has done. But a National spokeswoman says that her firm's research showed a limited market potential for the device. Eventually, Chesebrough-Pond's ended the agreement. A separate set of negotiations is going on with a controls manufacturer for the three-phase design.

A number of controller manufacturers—e.g., Panametrics Co. (Orange, Conn.) and Westinghouse Electric Corp. (Pittsburgh)—also offer the circuitry necessary for power-factor sensing as an add-on to their variable-speed or motor-control centers. Reliance Electric Co., the Euclid, Ohio, motor and controller manufacturer, has taken another tack: selling motors with built-in capacitors, an alternative method of power-factor correction. The company guarantees a minimum 0.95 power factor under full load.

A PROGRAMMABLE MOTOR—New electronic capabilities also are evident in the programmable motor announced last fall by GE. A year ago in England, a team from Leeds and Nottingham Universities that had been working on an electric automobile motor since 1975 arrived at a similar motor design, which it calls "switched reluctance." (Reluctance is a measure of resistance to magnetic flux. In motor parlance, it refers to ways of making the motor rotate.) This work by Peter Lawrenson and others is now being commercialized by Tasc Drives, whose U.S. distributor is CDL International (Clarence, N.Y.).

In a simplified definition, these motors are d.c. types, but lacking the brushes and commutator normally needed to regulate d.c. current. Instead, a sophisticated electronic circuit—in GE's case, featuring a type of metal-oxide semiconductor field-effect transistor (MOSFET)—regulates the current. (This year, MOSFET will be replaced by a newer device called an insulated gate transistor, which is cheaper.) The Tasc motor uses a set of thyristors (electronic switches) as the heart of the switching mechanism.

D.c. motors have the advantage of inherently variable speed (the current is simply reduced), but such drawbacks as: expensive construction in high-horsepower designs, and the increased maintenance requirements of the brushes and commutator. Moreover, the motors usually are not suitable for flammable chemical environments unless the sparking commutator is enclosed—a not-so-simple modification.

The new motors' electronic circuitry avoids these problems. GE notes that the units will cost more than conventional types, but that this cost should be offset by improved performance. The company says that industry could be selling more than 15 million such motors within ten years—a business that could be worth $700 million.

As yet, the GE motor is available only in fractional-horsepower sizes, for applications ranging from computer disk drives to ceiling fans. A company spokesman says that it will be at least a couple of years before larger sizes are sold.

Meanwhile, Tasc, which is already making 10- to 30-hp motors, and will soon begin manufacturing a range from 5 to 80 hp,‡ has published literature indicating that one of these motors costs roughly 50% less than a comparable a.c. motor with a variable-frequency drive. Joe O'Sullivan, managing director at Tasc, says that tests comparing the Tasc drive to the a.c.-motor/variable-frequency-controller package indicate a power saving for the former of 144,000 kWh/yr for a 200-hp fan.

NEW MAGNETS—On the horizon of motor design is the hope of replacing a substantial portion of motor windings with permanent magnets. The change would be significant because it would reduce power consumption, since the magnet would not be energized by current. But many questions remain concerning the adaptability of such magnets to industrial-size motor frames, and the ability to produce large magnets economically.

A new metal formulation for permanent magnets, which scientists in the U.S. and Japan are hurrying to commercialize, would contain neodymium, iron and boron (with the chemical formula of $Nd_2Fe_{12}B$). Apparently, GM and Sumitomo researchers have discovered that this mix of elements forms microstructures that can hold strong magnetic charges well.

GM reports that its material, called Magnaquench (because it is formed from liquid mixtures cooled in frac-

‡The motors are made for European voltages (380-415 V), so they would need transformers for U.S. use.

tions of a second), can achieve 30 million gauss-oersteds (MGO) in energy product (a technical term that denotes the magnet's strength); by comparison, conventional ferrite magnets are rated at 4 MGO. (Other specialty magnets have energy products in the range from 10 to 30 MGO, but have higher mass densities, and possess some performance constraints.)

Sumitomo says that it is "a step ahead" of GM in that it has constructed a 24-metric-ton/yr semicommercial plant to produce its magnet, called Neomax. The Japanese version has an energy product of 36 MGO, and Sumitomo says it has recently achieved 38 MGO.

A company spokesman notes that the lighter density of Neomax vs. that of other magnets makes it attractive for applications such as motors in automobiles (i.e., the starter), where weight savings are valued. (The specific gravity of Neomax is 7.4, compared with 8.4 for rare-earth/cobalt magnets.) But for industrial motors, "the key question will be the cost/performance ratio," he says, because the magnets are likely to be expensive (no cost data are yet available). He adds that the company foresees some shift over the next decade to the magnet, and is working with manufacturers to develop applications.

CPI STANDARD MOTOR?—While work on magnets occupies the Japanese, some groups in the U.S. are trying to develop a motor-design standard for CPI plant applications exclusively. The effort began several years ago when motor experts from a variety of companies realized that they had a list of criteria that would make it much easier to select chemical-duty motors.

"We thought we could get this standard established through the offices of the Chemical Manufacturers Assn. [the Washington, D.C., trade group]," says John Reynolds—a senior staff engineer at Union Carbide Corp.'s Charleston, W.Va., complex, who was involved in the early effort. "The standard went through several drafts, but then CMA decided against being a standards-setting body, and the effort was moved to a number of committees within the Institute of Electrical and Electronics Engineers [IEEE; Washington, D.C.]."

Currently, the matter is in the hands of a group called the RP-841 project (named for the eventual standard—the Recommended Practice, 841, for Chemical-Industry, Severe-Duty, Squirrel-Cage Induction Motors), and is sponsored by two committees within IEEE: the Power Engineering Soc. and the Industrial Applications Soc.

Andrew Smith, an engineering supervisor at Du Pont's Charlotte, N.C., plant, is the working-group chairman; he says that one draft last year drew extensive comments from user and manufacturer representatives; a new draft is now being completed. "We hope to have it out of our group and before the sponsoring societies by the end of the year, but it is premature to name a date when it will be complete and approved," he observes.

Traditionally, severe-duty motors in the CPI are of the totally-enclosed, fan-cooled type, and are often explosionproof as well. The proposed standard would go beyond these definitions in several key areas: housing (frame) construction, noise production, seal and bearing design, and insulation quality.

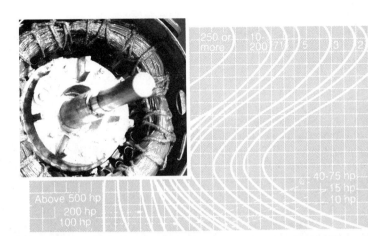

SPECIFYING ELECTRIC MOTORS

Edward J. Feldman, Siemens Energy & Automation, Inc.

Chemical engineers often get involved in specifying electric motors when a plant is being built or modified. This article explains how to choose an economical and reliable motor of the proper size and type, and discusses the workings of motor controllers (see p. 155).

Are you generating complete motor specifications? Some engineers describe what they need with just a few basic facts; e.g., 100-hp output, 1,800 rpm, 440-V three-phase power. The motor manufacturer asks for additional information, such as the operating environment and which way the shaft is supposed to turn, before selecting a particular motor. Another commmon practice, especially in a plant revamp or expansion, is to pull an existing specification out of the files and resubmit it.

Unfortunately, either approach can lead to omission of necessary features or inclusion of unnecessary ones. And reusing an old specification can lead to a totally wrong decision: What if the application or environment has changed? What if the old motor was specified incorrectly?

For all these reasons, the motor specification should be carefully tailored for each application. This article explains how to choose an electric motor that will economically and reliably deliver the required power. Fig. 1 is a cutaway of a typical horizontal motor, and its component parts; the enclosure is not shown. Table I provides an overview of what a specification should contain.

The main focus here is on alternating-current induction motors that are used as prime movers in chemical-process plants. Direct-current motors, and motors that are not used as prime movers, are a separate subject. Special attention is given to larger motors, of 500 hp or more, not covered by standards. Smaller motors are easier to specify because they are more or less standardized as to size, construction and torque.

In the U.S., the applicable standards are those of the National Electrical Manufacturers' Assn. (NEMA). Readers who need more information on motors are advised to look at the "NEMA Standards Publication for Motors and Generators," a thick volume that details the performance and construction of a.c. motors in the 1–450 hp range.

Torque requirements

An electric motor is essentially a torque generator; its purpose is to deliver the required torque through a shaft at a certain rotational speed. Therefore, the first thing to consider is how to match the the torque-generating capability of a motor with the load requirements of the driven device — pump, agitator, etc.

Torque, speed and shaft power are related as follows:

$$\text{Shaft power} = (\text{Torque})(\text{Speed})/(5{,}250)$$

where power is in hp, torque is in ft-lbf, and speed is in rpm. Thus, a motor that exerts 1,020 ft-lbf at 1,800 rpm is delivering 350 hp. Likewise, input (electrical) power can be calculated from the shaft power:

$$\text{Input power} = (746)(\text{Shaft power})/(\text{Efficiency})$$

Originally published May 11, 1987

where input power is in watts. A 350-hp motor having an efficiency of 0.85 draws 307,000 W, or 307 kW.

Voltage and current are related to the electrical power as follows. For a three-phase a.c. motor:

Input power = 1.732 × Voltage × Current × Power factor

A 350-hp motor having an efficiency of 0.85, as above, requires 307 kW of electrical power, and draws 458 A of 440-V three-phase current if the power factor is 0.88. For a single-phase a.c. motor, "1.732" (the square root of 3) is omitted from the equation.

The best way to evaluate torque requirements is to superimpose the speed-vs.-torque curve of the driven equipment on the speed-vs.-torque curve of the motor; these curves are provided by the respective manufacturers. NEMA has established minimum torque requirements for three basic motor designs: Design B, for normal centrifugal loads; Design C, for loads requiring high starting torque; and Design D, for high-slip high-inertia loads such as flywheel drives on machine tools. Within each class, minimum torque outputs are specified for each combination of nominal horsepower and operating speed. Fig. 2 shows the minimum-torque requirements for 1,800-rpm Design B motors. Fig. 3 compares a generic Design B motor with Design C and Design D motors; note that the Design D motor must be able to generate 280% of its full-load torque at startup.

Fig. 4 shows the speed-vs.-torque curve for a low-inertia load (such as a centrifugal pump) with a motor curve superimposed on it. Make sure that the speed-vs.-torque values are for the voltage applied to the motor terminals during acceleration. A motor that is being started across the line typically draws 650% of its full-load current, which reduces the voltage applied at the motor terminals to about 90% of the line voltage. Also, the line voltage will drop if other motors are being started at the same time. For these reasons, the motor curve must reflect the torque produced by the motor during the voltage dip. The most important torque points are:

1. Locked-rotor torque (Point A on Fig. 4). This is the torque generated by the motor when power is applied to the terminals and the rotor is still at rest. To accelerate properly at startup, the motor must generate more torque than the load requires. Otherwise, motor and load will remain at rest, and eventually the overload protection will trip the motor off the line. If the motor trips too late, it may be damaged.

2. Minimum accelerating torque (Point B). The danger of stalling is greatest when the difference between motor torque and load torque is at a minimum. To assure that the load is accelerated quickly enough, the specified motor should develop at least 30% more torque than the load requires at this point.

3. Maximum torque (Point C). The coupling between the motor shaft and the load shaft must be able to transmit more than the motor's maximum torque.

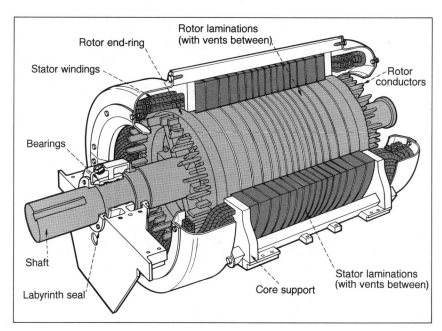

Figure 1 — Cutaway view of a horizontal induction motor

Table I — A complete motor specification covers electrical and mechanical components, in addition to speed, power and torque

Always specified	Specified when applicable
Type of enclosure	Vertical-motor shaft type
Altitude	Screens and filters
Ambient temperature	Vertical-motor rotation and thrust
Temperature rise	Space heaters
Insulation class	Bearing thermal detection
Starting method	Vibration detection
Mounting	Stator thermal detection
Inertia at motor shaft	Surge protection
Locked-rotor current limits	Current transformer
	Coupling to driven equipment
Vibration/balance tolerance	Base
	Grounding provisions
Efficiency	Drains
Terminal boxes	Testing to be performed
Noise level	
Shaft extension	
Direction of rotation	
Bearings	

Table II — Recommended voltages for electric motors

Rated voltage, V	Recommended power range, hp
230 or 460	Up to 100
460 or 575	100 – 600
2,300	200 – 4,000
4,000	400 – 7,000
6,600	1,000 – 12,000
13,200	3,500 – 25,000

Source: "NEMA Standards Publication for Motors and Generators," Section MG1-20.11, Part 20, p. 1.

SPECIFYING ELECTRIC MOTORS

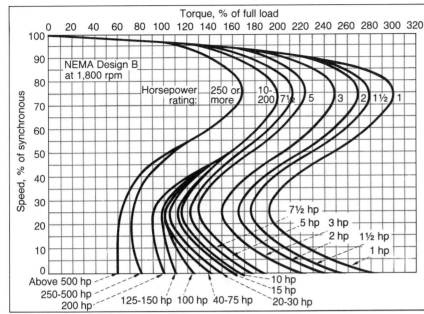

Figure 2 — Speed-vs.-torque curves define the minimum torque output of 1,800-rpm NEMA Design B motors

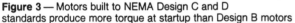
Figure 3 — Motors built to NEMA Design C and D standards produce more torque at startup than Design B motors

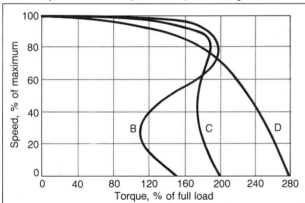

4. Operating point (Point D). The operating point is the intersection of the load and motor curves. The torque requirements of the load at this point should never exceed the rated full-load torque (i.e. 100%, in Fig. 4); otherwise, constant overloads will increase the operating temperature and shorten the motor's life.

So long as the motor generates more torque than the load requires, it will accelerate the load. When the motor output matches the load, the motor attempts to operate the load continuously at that speed. Thus, the motor must be selected so that the operating point is at or below the rated full-load torque, and at or above the rated full-load speed.

Voltage and power supply

After torque, the next specification to consider is the voltage to be applied. If possible, use the voltage available at the motor's location in the plant. If a supply line must be extended, it should be from the nearest source.

There are limits on the combination of power and voltage. High power and low voltage requires large conductors (i.e., thick copper wires) to accommodate the high current (current is proportional to power/voltage). Such conductors can be difficult or impossible to wind. On the other hand, low power and high voltage results in low current and small conductors; the wire may be so thin that it cannot mechanically support itself.

NEMA recognizes this problem and has recommended certain horsepower ranges for various voltages. Table II lists these ranges. In some cases, it may pay to specify an off-voltage motor (e.g., to take advantage of the available power supply). But before doing so, check whether such a motor is available. Also be aware that it will cost more and take longer to deliver.

Electric motors that use three-phase power are designed on the assumption that the applied line voltages will be balanced; i.e., that there will be the same voltage between Phases A and B as between Phases B and C and Phases C and A. If the phase voltages are unbalanced, the currents flowing through the three windings will also be unbalanced; a 1% imbalance in voltage can lead to an 8% imbalance in current. The result is that some parts of the stator will be heated more than others, producing hot spots that could damage the motor or cause it to trip. Further, unbalanced currents will try to turn the motor shaft in the wrong direction, reducing the net torque output.

NEMA standards allow 1% imbalance in the phase voltages. Manufacturers take this into account by designing motors to give satisfactory performance without shortening life expectancy as long as the imbalance remains within 1%. If line conditions are such that the 1% limit cannot be met, the motor must be oversized to accommodate the extra heating.

Motor-starting restrictions

The preferred method for motor starting is "across the line," which means that full voltage is applied to the terminals all at once. This provides the maximum torque for acceleration, and brings the motor and load up to speed in the shortest time. Unfortunately, across-the-line starting draws a high current (typically 6.5 times the full-load current), which may have a detrimental effect on other loads supplied by the same power source. The larger the motor, the more severe this problem becomes. If across-the-line starting is unacceptable, consider methods of reducing the starting current.

The most frequently used approach is to reduce the starting voltage. This can be accomplished by: a circuit that includes resistors and/or reactors in series with the motor winding; an autotransformer having adjustable-voltage taps; or a solid-state starter that adjusts the applied voltages to maintain a constant current. (For more information on solid-state starters, see the box on p. 155).

Another approach to reducing starting current is to use a wye-delta stator winding, in which both ends of each phase winding terminate in the conduit box. The starter connects the motor winding "in wye" on starting. This reduces the phase voltage by the square root of 3, thus reducing the starting current. Once the motor is up to speed, the starter reconnects the windings "in delta" for proper operation.

Smaller motors can be wound with two separate windings, only one of which is energized on startup. In this case, the starter's timing circuit must be adjusted so that the starting winding is not alone on the line for more than three seconds; otherwise, the overload can damage the motor.

As Table III shows, all reduced-voltage starting methods also reduce the torque developed by the motor during acceleration. Because there are limits to how long each type of starting can take, the acceleration requirements should be checked before specifying how the motor is to be started.

The formula for determining acceleration time is:

$$\text{Time} = \frac{(\text{Inertia})(\text{Change in speed})}{(308)(\text{Accelerating torque})}$$

where time is in seconds, inertia is the total inertia of motor and load in lbf-ft^2, change in speed is in rpm, and accelerating torque is in ft-lbf.

Since the accelerating torque changes as the motor speeds up (see Fig. 4), acceleration times should be calculated for small increments of speed and added up, or integrated numerically. To illustrate the method, however, we will use average values. For example: A 500-hp centrifugal pump having an inertia of 450 lbf-ft^2 is to be connected directly to a 1,775-rpm motor having a full-load torque of 1,479 ft-lbf and a rotor inertia of 142 lbf-ft^2. In accelerating from 0 to 1,775 rpm, the pump requires 45% of the motor's full-load torque, and the motor delivers 120% of its full-load torque, on average. Thus, the acceleration time is roughly:

$$\text{Time} = \frac{(450 + 142)(1,775 - 0)}{(308)(1.20 - 0.45)(1,479)} = 3.08 \text{ s}$$

In practice, it is important to make sure that the motor torque exceeds the load torque by at least 30% until the motor is nearly at full speed. Otherwise, the acceleration time-limit will expire, and the starter timer will take the motor off the line.

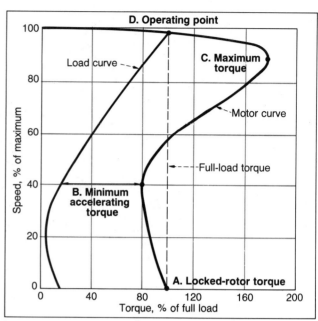

Figure 4 — The motor will operate at the point where its speed-vs.-torque curve intersects that of the load

Table III — Reduced-voltage starters also reduce the torque output of the motor

Type of starting	Relative starting current*	Relative starting torque	Relative smoothness of acceleration	Allowable acceleration time
Across the line	100%	100%	Smoothest	N.A.
Reduced voltage				
Resistor/reactor (at 65% voltage)	65%	65%	2nd smoothest	5–15 s
Autotransformer (at 65% voltage)	42%	42%	3rd smoothest	30 s
Wye-delta winding	33%	33%	4th smoothest	45–60 s
Two-part winding	50%	50%	Least smooth	2–3 s

*Compared with full-voltage across-the-line starting, which typically draws 6.5 times the full-load current.

Motor enclosures

The environment in which the motor is installed has a large effect on the life expectancy of the unit. While most motors are built to last for at least 20 years, their life expectancy will be severely shortened if they are not properly protected from dust, moisture and other contaminants. Fig. 5 illustrates the enclosures most commonly used for larger motors. The main difference among the enclosures is in the degree of protection given to the windings.

The open drip-proof (ODP) enclosure should be the first one considered since it is the least expensive. It is the basic unit offering a minimum of protection from the environment. An open motor uses outside air blown over the windings for cooling. A drip-proof enclosure is designed so that particles approaching the motor within 15 deg of vertical cannot enter the motor body or strike an inclined surface and roll into the interior.

If the motor is installed indoors, or at least under a roof, and the atmosphere is free of contaminants, the ODP motor will give long, troublefree operation; 15–20 years is common. The majority of motors have ODP enclosures.

If the environment is too wet for an ODP enclosure, a weather-protected Type I (WPI) enclosure should be considered. The WPI unit is essentially an ODP motor with a few additional features. Fig. 6 illustrates a long-shaft horizontal motor in a WPI enclosure.

In the WPI design, the insulation on the motor windings is more moisture-resistant — it may be coated, or made of a nonhygroscopic material. For large, high-voltage motors, form-wound sealed insulation is also available. The WPI enclosure includes screens over the air inlets and exhaust vents, to keep out pests and improve safety of operation. The

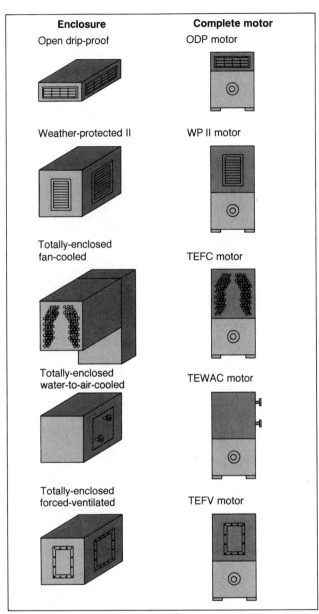

Figure 5 — The enclosure protects the motor from moisture and contaminants

Figure 6 — Horizontal motor with weather-protected (WPI) enclosure

rotor assembly has a moisture-resistant coating between the bearing fits for greater corrosion resistance.

The weather-protected Type I enclosure is recommended for indoor installations having small amounts of moisture in the atmosphere, or outdoor installations that have a roof or similar covering over the motor. A 500-hp motor (1,800 rpm, 2,300 V) having a WPI enclosure typically costs 10% more than a comparable ODP motor.

For outdoor installations and atmospheres containing dirt and dust, totally-enclosed fan-cooled (TEFC) or weather-protected Type II (WPII) enclosures are advisable. WPII motors are only available for sizes larger than NEMA 440, which is about 500 hp. TEFC is the choice for smaller motors.

The WPII is an open motor enclosure (using outside air for cooling) having moisture-resistant insulation (typically sealed) and screens. Additional features are shown in Fig. 7:

- Blow-through passageway. This allows high-pressure air to move through the enclosure without coming in contact with the rotor, stator and windings.
- Low-velocity chamber. Airspeed is held below 600 ft/min, which allows solid particles to settle out before contacting the vital internals.
- Baffles. The airstream must take three 90-deg turns, which helps settle out the remaining solid particles.
- Space heaters. These prevent moisture from condensing inside when the motor is shut down.

A 500-hp WPII motor costs about 65% more than a comparable ODP.

When the atmosphere contains acids, salts or other chemicals that could damage the windings, a totally enclosed motor should be specified. In such a motor, there is no free interchange of air between the inside and outside of the enclosure. Note that this does not mean an airtight seal; totally enclosed motors can and will breathe.

The most commonly used totally-enclosed motor is built with an external fan mounted directly on the shaft to blow air over the ribbed yokes on the motor body; this is the totally-enclosed fan-cooled type. Fig. 8 illustrates a vertical TEFC motor with water-cooled bearings. External fan cooling is not as efficient as having air blown directly on the windings. Thus, a TEFC motor must have a greater thermal mass — a larger stator and rotor, on a larger frame — to help dissipate heat. This makes the TEFC design more expensive than the others; a 500-hp TEFC motor costs about 90% more than a comparable ODP.

As motor size increases, heat-dissipation requirements also increase until the ribbed-yoke motor body cannot adequately cool itself. This situation calls for auxiliary cooling. One option is an air-to-air heat exchanger, mounted either above or around the stator. Most modern designs have the heat exchanger above the stator because that provides better cooling. Fig. 9 shows such an exchanger. Hot internal air circulated around the exchanger tubes is cooled by colder outside air blown through them. The tubes are usually made of aluminum, unless a different material is specified.

Because of the external fan, a TEFC motor will be noisier than an open motor — typically 3–6 dBA more. If this noise level is objectionable, consider a totally-enclosed water-to-air-cooled (TEWAC) design, which is available for motors of 500 hp and up. This is a totally enclosed motor that uses a water-to-air heat exchanger to remove heat. The exchanger is mounted above the stator, and air is blown over the water tubes by a fan mounted internally on the rotor.

Because water cooling is more efficient, a TEWAC motor can be built on the same size frame as an open motor. Still, it costs more — about 80% more than an ODP motor — and requires a supply of clean, cool water, most often from a cooling tower. Ideal conditions for cooling are (a) to use 80°F water of neutral pH and clean enough to keep the fouling factor at or below 0.001, and (b) to operate below 50-psi water pressure. The water requirement is about 0.03 gpm per rated horsepower. Because manufacturers can design the exchanger for other water conditions, water quality should be described in the specification. TEWAC motors should also be fitted with sealed, form-wound insulation.

If the motor is to be installed in an atmosphere where explosive gas, dust or fiber is usually present (National Electrical Code Div. I), the standard TEFC design is modified so that no spark or flame can escape should an explosion occur inside the enclosure, and so that the surface temperature of the motor cannot exceed 80% of the ignition temperature of gases and dusts in the air. In the 500-hp range, such an explosionproof motor costs about 105% more than a comparable ODP motor.

An explosionproof design is not called for in areas where explosive materials are not normally present but can be introduced into the atmosphere by a malfunction such as a broken pipe or valve (this is Div. II). For such areas, the specification is dictated by the normal atmospheric condition.

To obtain the correct motor for any atmosphere, it is necessary to specify the National Electrical Code Division, Class and Group that apply to the location. Within Div. I, Class I is for gases, including: acetylene (Group A); hydrogen (B); ethyl ether, ethylene and cyclopropane (C); and gasoline and petroleum products (D). Class II is for dusts: metal dust (Group E); carbon dust (F); or grain dust (G).

There is an upper size limit on the availability of labeled explosionproof motors. While this varies with individual manufacturers, and with speed and voltage, 1,750 hp is typically the largest size available. When a larger motor must be installed in an explosive atmosphere, a totally-enclosed forced-ventilated (TEFV) motor can be used. Here, the interior of the motor is pressurized so that contaminants cannot leak in. Clean air is ducted into the air inlets from an external source at a pressure of 1–2 in. of water above atmospheric. Internal fans move the air through the motor and into exhaust ducts that carry it out of the explosive area. TEFV motors cost about 75% more than ODP motors.

Insulation

In an electric motor, the stator leads, coils and connections are covered with dielectric insulation. The type used depends

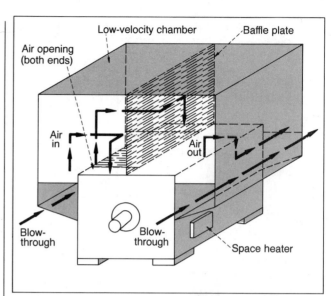

Figure 7 — This weather-protected (WPII) enclosure is designed to keep solid particles out of the motor

Figure 8 — Vertical motor with a totally-enclosed fan-cooled (TEFC) enclosure

on the service conditions, and on the voltage and size of the motor.

Most motors are designed for temperature conditions compatible with Class B insulation: a maximum internal operating temperature of 130°C. This means an allowed temperature rise of 90°C above ambient for an area (such as the continental U.S.) that has a 40°C maximum temperature.

Class F insulation, suitable for continuous operation to 155°C, is recommended for hot climates and high altitudes. If operated in a Class B environment, such insulation makes it possible to get more power from a motor, or to extend its useful life at normal power. Class H insulation, good to 180°C, is used for the same reasons as Class F is.

These types of insulation can be applied to random- or form-wound coils; Fig. 10 shows the two types of winding. Random-wound coils are built up of many strands of thin, round wire; each wire is coated with insulation. Due to the large number of turns, the voltage between adjacent wires is very low, and only a thin layer of insulation is needed. Because this approach is the least expensive, it is recommended for smaller motors operating at lower voltages, and is standard for voltages to 600 V and power ratings to 600 hp.

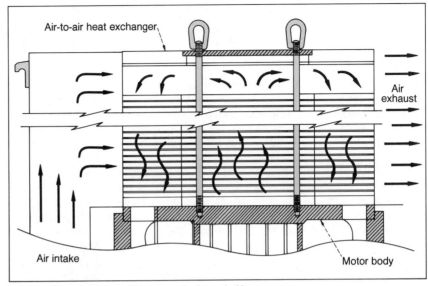

Figure 9 — Hot air from within the motor is cooled by outside air passing through the heat-exchanger tubes

The higher voltages used in large motors demand greater integrity from the insulation. Therefore, it is necessary to supply form-wound coils, though that is more expensive due to increased hand labor. Form-wound coils are made up of thick, flat wires shaped to fit the stator slots. Each coil is wrapped with sufficient insulating tape to form a dielectric barrier compatible with the system voltage. The wrapped coils are inserted into slots in the stator, and connected.

Most motor manufacturers offer vacuum-pressure-impregnated (VPI) coils, which should be considered for motors to be installed in petrochemical atmospheres. This Class F (155°C) system utilizes form-wound coils with mica tape as the insulator. The wrapped coils are inserted into stator slots and connected. Then the complete stator is immersed in epoxy resin and subjected to multiple vacuum and pressure cycles. This fills all the voids in the tape, and between the coils and slots. When the epoxy cures, the result is a very rigid and mechanically strong system that can withstand severe electrical and mechanical surges without harm. Because of the epoxy, the system also resists corrosion.

Temperature rise and service factor

Motors heat up as they run. To operate reliably for a long period of time, like 20 years, the maximum inside temperature must be kept below the limits of the insulation. This will normally be the case, so long as the motor operates at its rated power. However, changes in the load requirements can cause the motor to temporarily exceed its rated power. The motor will deliver the required torque, but its operating temperature will increase. If such load variations are likely, specify a motor having a service factor (SF) of 1.15. A 1.15-SF motor costs more, but it can safely operate for short periods at 15% above its rated horsepower.

The maximum operating temperature for Class B insulation is 130°C (266°F). Subtracting a maximum ambient temperature of 40°C (104°F) leaves a 90°C (162°F) limit on temperature rise. In practice, the motor temperature is not

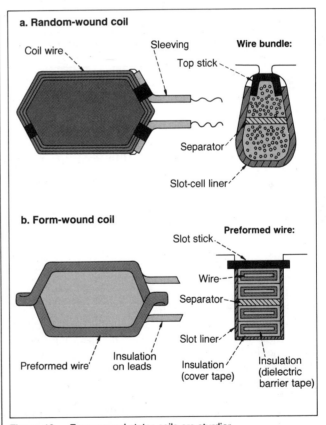

Figure 10 — Form-wound stator coils are sturdier and better insulated than random-wound ones

measured directly, but is calculated from the change in resistance of the winding. Since parts of the motor can be 10°C hotter than this average value, 80°C (144°F) is the practical limit for temperature rise in Class B motors.

This 80°C limit applies to normal Class B motors, which have an SF of 1.0. Motors having service factors of 1.15 are designed for a 90°C temperature rise (as measured by resistance), and can produce 15% more power. Typically, such motors will be operated at increased power only for short periods of time, and at rated power the rest of the time.

Table IV shows how operating a motor above or below its power and temperature rating affects its insulation life expectancy. Here, 20–25 yr is considered an average life expectancy for insulation. Thus, the life of a Class B motor having a 1.0 service factor will be reduced from 20–25 yr to 2–4 yr if it is continuously operated 15% above its rated power. Conversely, Class F insulation will have a longer life when operated at a lower Class B temperature rise.

Per NEMA standards, motors rated 1,500 hp and larger should be supplied with stator temperature sensors, either resistance temperature detectors or thermocouples. These devices are installed in the stator slots, between the coil sides, so that they measure the hottest part of the stator. Whereas the resistance method yields an average temperature, sensors indicate the hottest temperature. Therefore, a greater temperature-rise reading is expected when the motor is equipped with sensors: For motors having Class B insulation and an *SF* of 1.0, a 90°C temperature rise by sensor is standard to 1,500 hp. Beyond that, 85°C is standard.

Climate, altitude and inertia

At high altitude, the air is less dense and therefore cannot carry away as much heat. The effect on operating temperature is negligible to about 3,300 ft (1,000 m), but beyond that it must be accounted for. One option is to specify a class of insulation suitable for the higher operating temperature. The other choice is to specify a larger motor and operate it below its rated power. Recommended derating factors are: 0.97 for 3,300–5,000 ft; 0.94 for 5,000–6,600 ft; 0.90 for 6,600–8,300 ft; 0.86 for 8,300–9,900 ft; and 0.82 for 9,900–11,500 ft. Thus, a motor installed at 6,000 ft altitude should be run at about 94% of its rated horsepower.

The ambient temperature in which the motor is installed also has an effect on the size specification. In the continental U.S., manufacturers design their motors for an ambient temperature of no more than 40°C. If the ambient temperature is higher, as in a hot part of the plant (e.g., furnace room) or a hot desert climate, the operating temperature will be higher as well. To overcome this problem requires either a higher class of insulation or an oversized motor. What if the maximum ambient temperature is well below 40°C? In an arctic climate, the motor output can be increased.

The external load inertia is another factor that can effect motor size. The greater this external inertia, the more heat the motor will generate as it speeds up. NEMA specifies minimum values of external inertia that must be accommodated. If the actual external inertia is greater than these values (e.g., for a large fan or centrifuge), the motor manufacturer must examine the design to see whether it can stand the extra heat.

This heat increases the temperature of both the rotor and stator. In the rotor, the bars and end-rings are most effected. In the stator, the coils must be checked to determine whether the temperature will stay within the limits of the insulation. When design modifications are required, they will increase the initial price, but refusal to recognize them could result in downtime.

Table IV — Relative life expectancy of motor-insulation systems (average is 20–25 years)

Insulation class/ allowed temp. rise	Service factor = 1.0		Service factor = 1.15	
	Temp. rise, °C	Relative life	Temp. rise, °C	Relative life
Motors designed for service factor = 1.0				
Class B/80°C	80	1.0	105–115	0.1–0.16
Class F/105°C	105	1.0	140–150	0.1–0.16
Class F/105°C	80	6.0	105–115	0.5–1.00
Motors designed for service factor = 1.15				
Class B/80°C	65–70	2.0–3.0	90	0.5
Class F/105°C	85–95	2.0–3.0	115	0.5
Class F/105°C	65–70	12.0–16.0	90	3.0

Bearings and vibration

For horizontal motors, the most frequently supplied bearings are standard deep-groove ball bearings. These are the least expensive initially and the least costly to maintain, so they should be chosen except in unusual cases.

In NEMA frame sizes (up to 500 hp), ball bearings are the only type available for all speeds. Some motor manufacturers use open bearings, while others supply single- or double-shielded ones. The shields protect the rolling elements from contaminants, while still allowing for replenishment of grease. For severe environments, double-shielded bearings offer maximum protection.

All grease-lubricated ball bearings should be supplied with inside end-caps to prevent grease from migrating into the motor and to help keep internal contaminants out. The excess-grease reservoir should be located outboard of the bearing; and there should be a means of purging the old grease. To prevent overgreasing by the maintenance staff, the access and drain lines should be plugged rather than supplied with grease fittings.

Belt and chain drives exert greater radial forces than direct drives. Ball bearings have ample load capacity for low ratings and low speeds, but heavy radial loads at high ratings and speeds may require roller bearings. For this reason, specify details of the driven load as soon as possible.

Ball- and roller-type antifriction bearings are available in sizes to 75 mm for ratings to 500 hp and speeds to 3,600 rpm. If the application calls for more power or greater speed, the motor manufacturer must supply sleeve bearings — with tin-based babbited liners bonded to bronze or cast-iron shells.

Sleeve bearings cost more initially because they have more parts and require more labor to assemble, but in theory they can last forever if properly lubricated. They also vibrate less, and are quieter. Self-cooling bearings with ring-type oil

Electronic motor starters

Motors starting with a bang used to be the standard occurrence some years ago. Because there was no alternative to suddenly hitting the motor with full uncontrolled voltage, every start generated painful stresses that reduced the efficiency and life of the motor and the equipment connected to it.

The ideal way to start a three-phase motor is to gradually increase the voltage from a low value to the full line value, and to provide just enough current to generate the necessary accelerating torque. With electromechanical starters, the current and torque can be controlled only to a limited extent. Solid-state reduced-voltage starters, on the other hand, offer a greater degree of control over the current, speed and torque, and are able to start a motor smoothly while reducing power consumption. They can also control the rate of startup, and protect the motor against overloads and power fluctuations. Such starters are available for three-phase-a.c. induction motors having power ratings up to 350 hp and more.

How starting is controlled

The across-the-line starting method applies full line voltage to the motor terminals through a manual or magnetic starter. Unfortunately, the timing of contact closure is not controlled in electromechanical systems. The resulting phase imbalance can produce extremely large torque transients that can damage the motor. A solid-state reduced-voltage starter (SSRVS), on the other hand, applies the initial energizing voltage to the terminals at the ideal time and in the correct phase sequence, even when it is set to start the motor at 99% of the full line voltage.

After the initial energizing, the basic contactor starter has no control over the large inrushing currents, which can be eight times the full-load current in the case of high-efficiency motors. Starting at a reduced voltage lowers the maximum current, and for this reason is generally preferred.

Traditional reduced-voltage starting circuits make use of autotransformers, wye-delta winding connections, primary resistors and primary reactors (see p. 39 of the article). While these are more or less effective at reducing the startup current, all such electromechanical devices have limitations. For one thing, an excessively long time at a much-reduced voltage will lead to overheating of the starter, overheating of the motor, and possibly an overload-induced trip. To overcome this problem, the starter timing circuit must be adjusted by trial and error. Also, no mechanical device can turn the current on at the exact moment when the voltage passes through zero for all three phases, which is the best way to minimize torque transients.

A solid-state reduced-voltage starter, in contrast, can deliver the voltage to each phase at the right time, and can ramp smoothly from the initial voltage to full line voltage. Using its microcomputer to control the logic and timing, such a starter applies voltage by delaying the firing of the silicon-controlled rectifiers (SCRs). Because the starter gradually increases the voltage from a low level, torque transients are almost eliminated. As the motor speeds up, the starter controls both the current and the torque output.

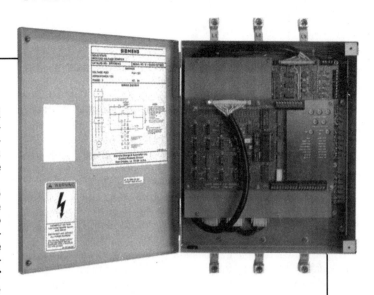

Such control is called "soft" or "stepless" starting. The major benefit is that it reduces the mechanical shock and vibration associated with conventional starting, and thus extends the life of the motor and the driven equipment.

Features of solid-state starters

Solid-state reduced-voltage starters use two electronic boards: The analog board contains current-sensor inputs and SCR trigger-pulse outputs, plus light-emitting-diode (LED) indicators. The digital-logic board contains: rotary switches for setting the full-load current (amps), percent voltage at start, ramp time, and current limit; on/off switches that enable remote starting, reversing and stopping; and LED displays of the controller status. The complete starter unit may include an enclosure, heat sink and protective devices.

When the operator pushes the start switch, the microcomputer verifies that the current setting and ramp time do not exceed the starter's rating. If they do, the start is locked out. Otherwise, the starter applies a small voltage to the motor and measures the resulting current for 10 cycles (1/6 s for 60-cycle a.c. current). Under normal conditions, this current is well below the motor rating; if not, the start is aborted. The microcomputer then computes the rate at which to ramp the voltage in order to reach full voltage at the desired time. As the starter fires the SCRs, it adjusts the timing for the smoothest possible starting.

Once the motor is up to full speed, the starter acts to protect it and to optimize energy usage. If the load on the motor changes, the starter adjusts the voltage to use only as much power as necessary. If the load is removed, the starter reduces the voltage as much as possible. Meanwhile, it continually checks the line, load and SCRs for malfunctions. When a fault occurs, the tripping time depends on the detected load: During starting, the starter trips the motor immediately if there is a phase reversal or SCR failure. At full load, the starter delays the trip, allowing the motor to ride out a temporary current imbalance or voltage dip.

Elie G. Ghawi, Siemens Energy & Automation, Inc. (Controls Div., 3333 State Bridge Rd., Alpharetta, GA 30201)

lubrication is the preferred construction; external-flood oil lubrication should be specified only when necessary. Oil-level indicators should be provided; constant-level oilers will ease maintenance. Pressure-equalizing vents installed inboard of the bearings will prevent oil from migrating into the windings. Because a sleeve bearing cannot handle any external thrust, it must be used for direct connection only. A limited-end float coupling will be needed to prevent the shaft from striking and damaging the side of the bearing.

In selecting bearings for vertical motors, the key factors are the thrust requirements of the load (i.e., how much vertical force it exerts) and the desired life expectancy. Usually, bearings are sized for the operating thrust and designed to last a minimum of one year; the method of calculating bearing life is Standard B-10 of the Anti Friction Bearing Manufacturers Assn. However, starting and stopping conditions must be carefully reviewed, particularly if momentary upthrusts or downthrusts can occur — e.g., when an agitator is started, or a pump is shut off.

Grease-lubricated deep-groove ball bearings can accommodate a small amount of thrust. Choices offering greater thrust or longer life expectancy are: angular-contact ball bearings, spherically-seated roller bearings, and finally tilting-pad thrust bearings. These options require oil lubrication and external water cooling.

Since the rotor is not a homogeneous mass, it will cause the motor to vibrate. NEMA recognizes this and sets maximum acceptable levels of vibration for standard motor designs. The amplitude of vibration (inches) is measured on the bearing housing at the horizontal, vertical and axial centerlines of the motor. Fig. 11 shows where the measurements are taken, and lists the maximum amplitudes for different speeds (any size motor shaft). These amplitudes should be specified unless the driven device has other requirements.

For instance, the vibration limits for high-speed compressors and pumps having mechanical seals are sometimes specified for the motor shaft rather than the bearing housing. This is to reduce seal wear in the driven equipment. Such shaft-vibration limits are usually much tighter than the NEMA standards, and therefore require special designs and manufacturing procedures — e.g., more-precise balancing, closer tolerances, use of sleeve bearings. Obviously, this increases the initial cost of the motor.

Efficiency

So far, we have considered what is involved in specifying an electric motor that can reliably start and drive a particular load. To assure a reasonable initial cost, the general rule is to specify no more or less than what is necessary. However, operating cost is also an important concern, and efficiency is among the key factors in determining what that will be.

Electric motors are by far the most efficient prime movers available: They convert approximately 85–97% of their input (electric) power into shaft (mechanical) power, vs. 40–50% for hydraulic motors, and 30% or less for steam and gasoline engines. Even when hydraulic energy and steam are less expensive than electricity, the electric motor is generally more economical. Still, motor efficiency varies significantly among designs and manufacturers, and must be considered a key part of the motor specification.

If a motor is to be used infrequently, additional investment in a more-efficient design is not warranted. Choose the least expensive design that meets the requirements, regardless of its efficiency. On the other hand, if the motor is to be run continuously for thousands of hours per year, a more-efficient design will quickly pay back its additional cost.

A motor's efficiency is improved by cutting the energy losses due to electrical resistance and mechanical friction:

$$\text{Efficiency} = \frac{\text{Output power}}{\text{Input power}} = \frac{\text{Output power}}{\text{Output power} + \text{Losses}}$$

Ways of improving efficiency include:
- Using more copper in the stator windings, and thinner laminations of a better grade of steel.
- Changing the rotor material to copper, or increasing its copper content.
- Using lower-friction bearings, more-efficient fans, and larger motor frames to reduce the operating temperature.

All of these improvements cost money, but if the additional cost can be recovered quickly, they are worthwhile. To calculate the payback, consider at least:

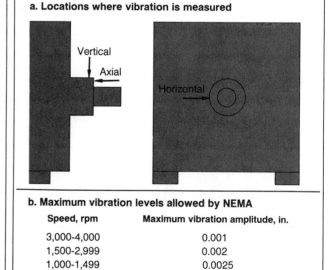

a. Locations where vibration is measured

b. Maximum vibration levels allowed by NEMA

Speed, rpm	Maximum vibration amplitude, in.
3,000-4,000	0.001
1,500-2,999	0.002
1,000-1,499	0.0025
Up to 999	0.003

Figure 11 — Vibration is normally measured at the bearing housing

- Extra cost of high-efficiency motor.
- Cost of power per kWh, now and in the future.
- Hours of operation per year.
- Reduction in power demand, kW, and power consumption per year, kWh.
- Tax rates and depreciation schedules.

Noise

At this point, it is necessary to consider how the motor will fit into the plant environment: Is there room to install it? Will it strain the electric-power grid? Will it make too much noise?

Noise is important because it affects the quality of the work environment, and is subject to government regulations; it is the user's responsibility to make sure that noise regulations are met. In the U.S., the Occupational Safety and Health Administration (OSHA) permits no more than 90 dBA total noise at locations where employees will be stationed for eight hours at a time. If the motor is the only piece of equipment, its noise must be no greater than 90 dBA when measured at the appropriate distance. If other equipment will be operated nearby, the motor may have to be quieter.

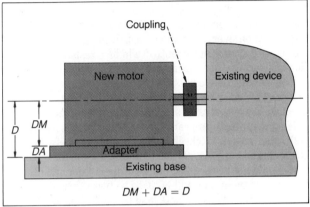

Figure 12 — Dimensional specifications are critical when fitting a new motor to an existing device

Motor noise is associated with the movement of air. It can be reduced by installing baffles and lining in the motor's air passageways; of course, these features increase the cost. The motor housing also helps to muffle noise. In general, open drip-proof and weather-protected motors are quieter than totally-enclosed fan-cooled (TEFC) ones. If TEFC enclosure is not required, noise can be reduced by choosing one of the other enclosures.

If noise requirements are not specified, the manufacturer will provide a standard design. For a 500-hp, 3,600-rpm, 460-V motor, noise levels measured 1 meter away are typically:

Enclosure	NEMA maximum	Standard design	Low-noise design
ODP	99 dBA	90 dBA	82 dBA
WPII	94	85	76
TEFC	103	91	81

Dimensional limitations

However efficient and reliable a motor may be, it is not a suitable choice if it cannot be moved into place and properly connected to its load. For this reason, the job site should be checked for obstructions such as overhead pipes, columns, and doorways, and dimensional limits should be specified.

Obtaining a motor that can easily be connected to the driven shaft is a key concern when replacing an existing motor. The most critical dimension is the distance from the shaft centerline to the mounting feet — dimension *DM* in Fig. 12 — because for proper coupling the replacement motor shaft must match the driven shaft within about 0.015 in.

Unless the motor being replaced is a very recent design, it is improbable that an exact dimensional duplicate will be available, because modern motors are generally smaller and shorter than older ones. The more economical approach is to purchase a modern design, and correct for the lower shaft height by installing an adapter plate of the appropriate thickness between the existing base and the new motor. Note that the coupling may also have to be changed if the new motor shaft is smaller in diameter than the old one was.

The location of the conduit box on the motor housing is best left to the manufacturer. For a replacement motor, however, the box must be positioned so it can be connected to the existing electrical conduit. Note that NEMA prohibits having high- and low-voltage conductors terminating in the same box. If auxiliary devices require a different voltage, a separate box will be required.

Power factor

Alternating electrical current can be thought of as a vector having two components at right angles: The current in phase with the voltage is capable of doing useful work. The right-angle current is out of phase; it flows through the motor and generates heat but produces no useful work. When the two components are added up, the cosine of the angle between the current vector and the voltage vector is the power factor.

With a pure-resistance load such as an electric light, the current is always in phase with the voltage. There is no right-angle component, and thus the power factor is 1.0. With the induction motors we have been discussing, the power factor is always less than 1.0: The current always lags behind the voltage because the motor coils have a certain inductance. The greater the inductance, the lower the power factor.

A premium for low-power-factor loads is charged by electric utilities because they have to deliver additional useless current. If the motor is going to upset the power grid, it may be necessary to install capacitors on the motor terminals to counteract the current lag. The amount of capacitance (measured in kVAR, or kilovolt-amps reactive) can be calculated by the motor manufacturer. Usually, such capacitors are connected so they can be switched on and off by the motor starter.

If the system power factor is already low, selecting a synchronous motor having a leading power factor (greater than 1.0) will help improve the situation.

Special requirements for vertical motors

So far, we have focused mainly on horizontal motors. Vertical ones are different in some respects, and these differences must be considered in the specification.

Vertical motors are used to drive equipment such as turbine pumps and liquid agitators. They are built on a round frame, and flange-mounted to the driven device. There are three standard NEMA flanges: The C flange is for driven equipment supported by the motor. Holes for mounting bolts are drilled and tapped into the motor housing. The D flange is for motors mounted on the driven equipment. Mounting holes are drilled in the equipment. The P flange is for vertical mounting of centrifugal pumps. The mounting holes are in the pump base.

Most vertical pumps call for motors with solid shafts for external coupling, but deep-well turbine pumps require a hollow-shaft motor so that the pump shaft can pass through the center of the motor shaft and be coupled to it at the top. This allows the pump clearance to be adjusted for wear.

In a solid-shaft vertical motor, the bearings are designed

to support the weight of the rotor. In a hollow-shaft design, the bearings must support the pump rotor as well, and must also be able to accommodate the hydraulic thrust load imposed by the pump. One year (per the B-10 method) is considered a minimum expected life for the bearings.

Protective devices

Reliable as electric motors are, they can be damaged when subjected to overloads and surges, and when bearings or insulation burn out. Fortunately, protective devices that will take the motor off the line before it is damaged are widely available. What kind of protection is needed depends on how often a load or power problem is expected to occur, and how important the motor is. If problems are likely and the motor is critical, it may be necessary to monitor the motor daily. Often, company policies dictate that monitoring and protective devices be installed.

Because they are vulnerable to overheating, the stator windings are the first thing to consider protecting. Resistance temperature detectors (RTDs) and thermocouples are used to monitor stator temperature. If there is no need for monitoring, a more economical approach is to install bimetallic switches on the end-turns of the windings. If the windings get too hot, as from a continuous overload, the switches will open and the relay will take the motor off the line.

Bearing-temperature monitoring should also be considered for large motors having sleeve bearings, in case the oil level drops too low to lubricate effectively. Either RTDs or thermocouples can be used for this purpose. Dial-type thermometers that give direct readings at the bearings are also available, but the operator must keep an eye on them. Automatic protection is provided by bearing relays that switch the motor off when a preset temperature is reached.

In most ODP, WPI and WPII enclosures, filters can be installed to capture large solid particles in the incoming air. Such filters are inexpensive and can be removed for cleaning. If they are used, it is advisable to monitor their performance to make sure they do not plug up. This can be done with a manometer or differential-pressure switch installed so as to measure the pressure on both sides of the filter. Stator-temperature detectors are also useful in spotting filter problems: As the filter begins to plug, the stator temperature begins to rise, indicating that the filter should be cleaned.

Protection against excessive vibration can be provided by a seismic-type vibration monitor installed on the outside of the motor housing. If vibration starts to increase, the motor needs maintenance. For sleeve-bearing motors, shaft vibration can be monitored using proximity probes. These generate a small magnetic field around the spinning motor shaft. As the center of mass of the shaft moves, the magnetic field changes in strength and shape. The probes measure this effect, and transfer the data to a proximeter that indicates the amplitude and frequency of vibration.

Vibration monitoring provides early warning of mechanical failure: If maintenance is started when a pattern of increasing amplitude is first detected, the motor will be off the line a much shorter time than if maintenance is delayed.

Any such probes (or provisions for mounting them) should be included in the specification and installed by the motor manufacturer. Trying to add probes later on will be more complicated and expensive; e.g., because the motor shaft under a proximity vibration probe must be polished.

Testing requirements

The final item in the specification is how the motor is to be tested. Again, there are choices, and some of them involve additional expense.

Electric motors are engineered products that need to be tested before they are shipped. Motor manufacturers perform standard commercial tests (using methods laid out by the IEEE) on every motor: no-load running test; insulation dielectric test; measurement of winding resistance, inspection of bearings, and vibration check. The numerical results can be made available but must be requested before the tests are performed. Otherwise, the test items may just be checked to see that they are within specified limits.

Most manufacturers can also load-test their motors at the full rated power to check the operating temperature. Because this is costly ($1,500–6,000) and time-consuming, such a test should not be specified unless there is a definite need for it — e.g., if the motor will be used to test other equipment, such as pumps. The complete load test will measure the full-load operating temperature, percent-slip curve, locked-rotor current, breakdown torque, starting torque, efficiency, power factor and winding resistance. In a critical case, the purchaser can witness the testing. Unfortunately, this doubles the cost, because the manufacturer will test the motor privately before advising that a test may be witnessed.

A final checklist

When you are writing a motor specification or providing data to the manufacturer, keep these few rules of thumb in mind:
• Be sure that the motor will produce sufficient torque to start and drive the load.
• Specify a NEMA-standard motor whenever possible.
• Choose an enclosure compatible with the environment.
• Specify vacuum-pressure-impregnated (VPI) insulation wherever practical, and especially in chemical environments.
• Include altitude, ambient temperature and external inertia in the specification if they are outside the usual ranges.
• Allow the motor manufacturer to select the bearings unless there is good reason to specify a particular type.
• Demand a high efficiency for any motor that is to be run continuously.
• Select a quieter motor whenever possible.
• Specify only the protective devices actually needed.
• Accept the manufacturer's routine tests.

In general, do not overengineer a motor specification. Electric motor designs are proven and reliable, and there is little to be gained by specifying unnecessary features.

The author

Edward J. Feldman is a senior application engineer for Siemens Energy & Automation, Inc., Large Motor & Generator Div., P.O. Box 25001, Bradenton, FL 33506, where his primary responsibility is in motors for the petrochemical industry. He is also involved with heating, ventilation and air-conditioning applications, and with training people in the design and application of motors; he has had over thirty years of experience in the field. He has a bachelor's degree in electrical engineering from the University of Cincinnati.

Check pump performance from motor data
V. Ganapathy*

Plant engineers often need to figure the flow or head of pumps or fans. Such information is needed to check performance or to see if meter readings are accurate; for meters, the flow orifice might be plugged or the meter might be incorrectly calibrated.

Motor data

Electrical-current readings provide a good indication of the flow. The following analysis is developed for pumps, but it may also be extended to fans, since they have similar characteristics.

The power consumed by a pump is:
$$B_{HP} = Q\Delta P/(1{,}715 E_p) \qquad (1)$$

The head may be related to the differential pressure by:
$$H = 144\Delta P/\rho \qquad (2)$$

Converting Eq. (1) into kilowatts:
$$P = 0.00043 Q\Delta P/E_p \qquad (3)$$

The power output of the motor is given by:
$$P = 0.001732 E I \cos\phi E_m \qquad (4)$$

Equating Eqs. (3) and (4):
$$Q\Delta P/E_p = QH\rho/144 E_p = 4.03 E I \cos\phi E_m \qquad (5)$$

Thus, pump data can be related to motor data.

Comments

The following should be kept in mind:

1. Motor efficiency does not vary much with load. The variation is probably 2–5% for medium-sized motors (50–500 hp) and for practical purposes can be considered constant. So can the power factor and voltage.

*ABCO Industries, Inc., 2675 E. Highway 80, Abilene, TX 79604.

Nomenclature

B_{HP}	Brake horsepower, hp
E	Voltage
E_p, E_m	Efficiency of pump and motor, fraction
H	Head, ft
I	Current, amps
P	Power consumed, kW
Q	Flowrate, gal/min

Greek letters

ΔP	Pump differential, psi
ρ	Density, lb/ft^3
$\cos\phi$	Power factor

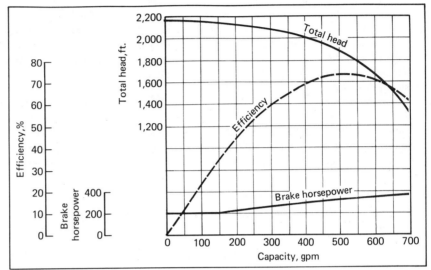

Figure 1 — Typical characteristic curve for a multistage centrifugal pump used to prepare table

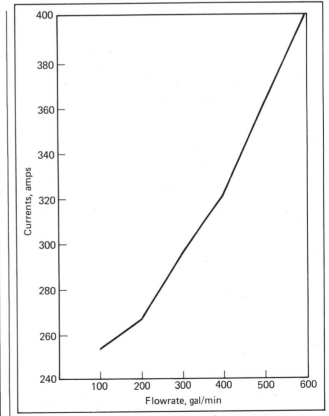

Figure 2 — Using motor data, this plot can be constructed and then used to check pump and meter performance

Table — Typical operating data for a pump

Q, gpm	H, ft	E_p	QH/E_p	I, amp
100	2,150	0.23	935,000	254
200	2,130	0.43	991,000	269
300	2,100	0.58	1,090,000	296
400	2,000	0.68	1,180,000	321
500	1,880	0.71	1,330,000	361
600	1,690	0.70	1,450,000	394

2. The head, H, and efficiency, E_p, of the pump vary parabolically — see Fig. 1 for a typical pump curve. For a given fluid and flowrate, $QH\rho/144E_p$ is a curvilinear function of Q. Thus, Eq. (5) may be written as:

$$f(Q) = f(I) \quad (6)$$

For a given pump and motor, if E, $\cos\phi$ and E_m are known, Q can be determined as a function of current.

For the pump curve in Fig. 1, we shall determine Eq. (6). First, a table is made of Q, H, and E_p. QH/E_p is calculated (see the table). Assume $E = 460$ V, $E_m = 0.95$, $\cos\phi = 0.9$ and $\rho = 62$ lb/ft^3. Substituting these data in Eq. (5):

$$QH/E_p = 4.03 \times 460 \times 0.9 \times 0.95 \times 144/62 = 3{,}680 I \quad (7)$$

By plugging in values from the table, I can be found as a function of Q. This is shown in the table and plotted in Fig. 2.

Using Fig. 2, one can easily relate the flowrate (or head or pump efficiency) to motor current. This is a convenient way of checking the performance of a pumping system or of double-checking meter readings. Thus, having a relationship between flow and current permits monitoring of any significant deviations. For example: What if at some flowrate, the meter shows a current reading that differs from that given by the curve for the same flowrate? Then either the flowmeter has to be checked (if the variation is greater than about 5%) or the pump has to checked for misalignment, overheating, etc. (The same method may, of course, be used on fans.)

The equations take into account friction and heat losses (of both the pump and motor).

Making the proper choice of adjustable-speed drives

Thomas R. Doll, Reliance Electric Co.

☐ New concepts to reduce energy consumption are being adopted in process plants. Rising energy costs have, for instance, resulted in stronger emphasis on the efficiency of drive systems.

In the past, the flow of process fluids was generally regulated by throttling, whereby a pump is operated at constant (full) speed and flow is varied by restricting it with a control valve. But this wastes energy.

Many centrifugal pumps, compressors, blowers and fans in process plants have fluctuating load requirements, yet their drives are sized for peak demand, even though (as Fig. 1 shows) that demand occurs for only a small portion of total operating time. Controlling flow with a valve, dampers, vanes or slip couplings is similar to driving a car at full throttle and regulating its speed by dragging the brakes or slipping the clutch.

Process control loops are increasingly being controlled by adjustable-speed drives, particularly by solid-state a.c. drives, because this type offers the ability to control the rate of energy consumption in the prime mover (see box below) and operates safely in hazardous atmospheres. Additionally, it can react to a variety of sensing devices that can change its speed in proportion to sensor signals derived from such variables as temperature, pressure, level, density or viscosity.

Basic types of adjustable-speed drives

Applications for adjustable-speed drives in process plants are wide-ranging. Many types of pumps (centrifugal, positive-displacement, screw, etc.) and fans (air-cooler, cooling-tower, heating and ventilating, etc.), as well as mixers, conveyors, dryers, calenders, crushers, grinders, certain types of compressors and blowers, agitators, and extruders, are driven at varying speeds by adjustable-speed drives.

Most of the adjustable-speed drives in process plants are rated for less than 500 hp. Within this group, five types prevail: solid-state a.c., solid-state d.c., mechanical, electromechanical slip, and fluid.

Because a.c. and d.c. drives function by altering the operating speed of the prime mover, they are preferred when energy conservation is a prime consideration.

Flow and power relationship: key to saving energy

Flow volume from centrifugal pumps and fans is related exponentially to motor horsepower.

The first graph below shows that the relationship between flowrate and motor speed is linear; when more flow is required, a proportional increase in motor speed will deliver it.

However, the second graph indicates that line pressure increases with the square of motor speed. And the third graph says that motor power requirement increases with the cube of motor speed.

The table demonstrates how dramatically power requirement is reduced when flowrate is lowered. For example, cutting the flowrate by 20% reduces motor speed proportionally—but decreases the power requirement by 49%!

Speed, %	Flow, %	Horsepower required, %
100	100	100
90	90	73
80	80	51
70	70	34
60	60	22
50	50	13
40	40	6
30	30	3

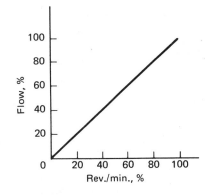 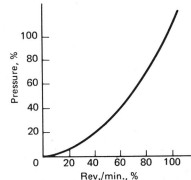 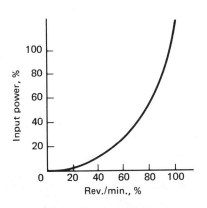

Originally published August 9, 1982

Performance and operating characteristics for adjustable-speed

System:	Solid state		Electromechanical	
Type of drive:	Electric, a.c.	Electric, d.c.	Eddy-current clutch	Wound-rotor motor
Maximum power, hp	500+	500+	500+	500+
Largest reduction ratio	>10:1	Infinite	5:1	Infinite
Speed regulation, %	0.5	1.0	3 to 5	2 to 5
Overall efficiency*,†				
Constant torque	9	9	8	8
Variable torque	8	8	5	5
Reliability*	9	9	7	8
Ease of maintenance*	9	9	7	7
Complexity††	10	8	7	7
How controlled	Open loop, remote operator	Tachometer feedback, remote operator	Closed loop	Closed loop, manual
Operating characteristics	Minimum maintenance, excellent system efficiency	Good low-speed response, precise positioning, moderate maintenance	Variable torque capability	Stable down to 50% of rated speed
Applications	Where maintenance is difficult, and energy is expensive	Where excellent speed control over a wide range is required, and sparking is not a hazard	Fans	Large pumps, high inertial loads

* Rating scale: 10 = best, 1 = worst.
†† Smaller numbers are more desirable in this category.
† For other than electric drives, overall efficiency includes the prime mover, an induction motor.

However, the other types of adjustable-speed drives have qualities that make them more suitable for certain applications.

Simple and inexpensive, mechanical-belt drives provide smooth operation and are capable of absorbing considerable shock loads. In addition, maintenance is relatively simple. Within a limited range, belt drives can operate at continuous reduction ratios. Being lightweight, they are frequently used with mobile equipment, such as portable mixers.

For applications that require precise, quick changes in speed, electromechanical clutches work well. The control mechanisms for these drives are highly adaptable to process-related inputs. By varying slip, electromechanical drives can provide indirect control of several variables, such as speed, position and power.

The wound-rotor motor is similar to the a.c. induction motor, except that its rotor has windings that are joined to three slip rings. Externally controlling the resistance in the rotor/slip-ring circuits enables the motor to operate as an adjustable-speed drive. Increasing this resistance reduces motor speed, because the energy directed through the resistors is converted into heat, which is dissipated as "slip loss."

Eddy-current clutches are by far the most common of the adjustable-speed electromechanical drives. They offer precise torque control, along with long service life in coupled (not belted) applications.

Hydroviscous drives are ideal in applications that must be run continuously at high horsepower levels. They can operate in environments where temperature varies widely and abrasive particles are present. Another advantage of hydroviscous drives (and all types of fluid drives) is their inherent safeness. Because torque is transmitted through fluid, there are no sliding parts to produce sparks, and operation is very smooth.

Load-factor criterion

The procedure for selecting the ideal drive for a particular application is complex. Not only must rugged-

Capacities of fixed drives are often underused   Fig. 1

PROPER CHOICE OF ADJUSTABLE-SPEED DRIVES

drives provide guidelines for selecting a suitable unit

	Mechanical			Fluid	
Rubber belt	Metal chain	Wood block	Hydrodynamic	Hydroviscous
100	100	20	500+	500+
10:1	6:1	12:1	3:1	20:1
2 to 5	0.5 to 2	3 to 5	3.5#	3
4	8	4	7	7
6	7	7	5	5
7	6	6	8	9
9	6	8	8	9
1	3	2	8	9
Pneumatic, manual, electric and vernier screw	Manual, hydraulic, vernier screw	Vernier screw	Manual or remote via scoop-tube angle	Mechanically varying distance between driving plates
Overload and jam protection	Compact, no overload protection	Excellent jam and overload protection, high torque	Poor low-speed efficiency; good for high, vertical loads	Smooth transition through speed changes
Conveyors, pumps	Permanent conveyors, fans and pumps	Grinders, mixers	Gear motors, air compressors, ball mills, conveyors, crushers, separators	Slurry, pipeline and shipboard pumps, large fans and conveyors

#Regulation is very poor at low speeds.

ness, control flexibility, efficiency, initial cost, service life and environment be considered, but also the type of load.

Most applications fall into the variable torque category; centrifugal pumps and fans are examples. Torque increases as the square of speed (see second graph in box on p. 47). The drives generally chosen for these applications are mechanical and electrical types.

Such equipment as screw pumps, sludge pumps, conveyors, and extruders require constant torque from a drive to maintain constant output. In such cases, the selection procedure is much more complex, and must consider the capabilities of the motor to start the high frictional load. Generally, electrical, fluid and electro-mechanical-slip drives are preferred for this type of load.

Solid-state a.c. drives

Solid-state drives are made up of a motor and a controller that processes the line power in such a way as to vary the rotational speed of the motor shaft to suit operational requirements.

Two basic types are available: alternating current and direct current. Today, the largest number of adjustable-speed drives in industry overall are d.c. Until recently, a.c. types were not competitive in cost with other types, particularly d.c. ones, because of the complex technology involved with varying the speed of an a.c. motor. However, developments in the past few years have resulted in significant cost reductions and, therefore, new interest in a.c. drives (see box, p. 50).

Although variable-frequency controllers are complex, a.c. motors are not, and it is this basic simplicity of motors that tempted designers to improve the performance of a.c. control systems.

The a.c. motor is lighter, smaller, more rugged, less expensive, and much more readily available than comparable d.c. motors. It also has no brushes or commutator to wear out or to provide a source of sparks. Finally, huge advances in efficiency over the past two years have made a.c. motors even more desirable. Small motors now operate near or above 90% efficiency; large motors operate at higher than 96%.

New generation of a.c. controllers

Advances in a.c. drives coincided with the development of the solid-state switch, particularly the silicon-controlled rectifier (SCR), which is still used in larger a.c. drives. As good as they are, however, conventional SCRs have not been the perfect solution to the complex circuitry of a.c. controllers (see box, p. 51). The problem is that SCRs introduce additional complexity; once turned on, they must be periodically switched off by what is known as a commutation circuit.

Last year, however, the development of a.c. controllers took another significant step forward. A new generation of controllers was introduced based on large power transistors (460 V) instead of SCRs. Transistors have the advantage of not requiring a bulky commutation circuit. Thus, the new controllers are simpler and

Saving kilowatts

Much of the recent interest in a.c. drives is because they offer a huge potential for saving energy. The reason is simple: There are far more a.c. motors driving pumps, fans, compressors, conveyors, centrifuges, mixers, separators, crushers and other equipment than any other type of prime mover. Today, the vast majority of these motors run at their base (constant) speed only, even when they do not have to.

By reducing motor speed during periods of low process demands, engineers are able to save substantial amounts of energy. Although d.c. drives can also save much energy, there are far fewer d.c. motors in the chemical process industries than a.c. motors, and d.c. drives are less suitable to the wide variety of applications in which a.c. drives are found. This ability to conserve energy is also an advantage that a.c. drives have over their mechanical, electromechanical and fluid counterparts.

Most a.c. adjustable-speed drives operate at an overall efficiency of about 90%, over a speed range that varies from half to full base speed.

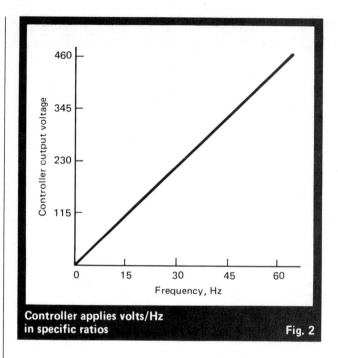

Controller applies volts/Hz in specific ratios — Fig. 2

more reliable, as well as much smaller and less expensive than those based on SCRs.

Another recent development in a.c. controller technology that has simplified SCR-type systems is the gate turnoff (GTO) switch. This is an SCR, but it is turned off by a negative signal at the gate terminal, instead of requiring a commutation circuit to interrupt signal flow.

Sizing a.c. controllers

The single most important factor in choosing an a.c. drive is the maximum current—for continuous and short duration—that the drive must handle. High starting torques require high starting currents, which may exceed the limits of the controller even though it mathematically can manage the constant-speed power needs of the application.

The key value that must be known to size a controller is the full-load current at base speed. This is the current requirement of a suitably sized motor operating under the anticipated load conditions.

How a.c. controllers work

Most solid-state adjustable-speed a.c. controllers used with standard induction motors produce variable voltage and frequency to control the motors. Frequency is controlled in order to vary the speed of the motor:

$$\text{Speed} \propto (K \times \text{Frequency})/N$$

Here, $K = 120$, and $N =$ number of magnetic poles.

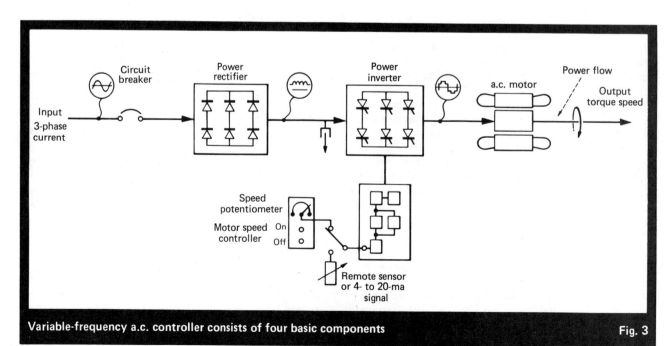

Variable-frequency a.c. controller consists of four basic components — Fig. 3

PROPER CHOICE OF ADJUSTABLE-SPEED DRIVES

Voltage is varied along with frequency so that the flux density in the air gap between the rotor and stator, and therefore the torque produced by the motor, can be controlled:

$$\text{Torque} \propto \Phi_{air\ gap}$$
$$\Phi_{air\ gap} \propto \text{volts/Hertz}$$

Here, $\Phi_{air\ gap}$ = magnetic flux density.

Typically, a constant relationship between voltage and frequency (volts per Hertz) is maintained (Fig. 2).

The basic components of a variable-frequency a.c. controller are a power converter, power inverter, control regulator and reference section (Fig. 3). The power converter changes line a.c. to d.c. The power inverter alters the d.c. to variable-voltage/variable-frequency a.c. power. The regulator controls the actions and response of the converter and inverter. The reference section is essentially a potentiometer and on/off switch, which tells the regulator when to turn on or off, and what speed is required.

Basic types of controllers

Three fundamental types of adjustable-frequency controllers are available up to 500 hp. Each features a different technique for converting line a.c. to d.c., then varying the d.c. to approximate a.c. Each has its own operational advantages.

The *variable-voltage-input* (VVI) inverter drive (Fig. 4) uses a controlled rectifier, or diode rectifier-chopper (not shown), to transform the incoming a.c. voltage into variable-voltage d.c. Frequency of the output is controlled by sequentially switching the transistors or thyristors in the inverter section in six discrete steps to produce the waveform output shown. The current follows the voltage in an approximately sinusoidal wave.

The VVI controller has the simplest regulation scheme of the three types of variable-frequency drives, but it uses the largest amount of d.c. filter components. These consist of the d.c. inductor and filter capacitors that filter voltage going into the inverter section and store energy for transient use.

The *current-source-input* (CSI) inverter drive (Fig. 4) also uses a controlled rectifier, or diode rectifier-chopper (not shown), to convert a.c. into variable potential d.c. Current sensed through transformers on the a.c. line serves as the basis for varying the controlled rectifier. The inverter section produces variable-frequency six-step current, and the voltage follows the current (with commutation spikes due to thyristor firing, as shown).

The main advantage of the CSI drive is its ability to completely control motor current, which results in complete torque control. However, this current-controlling characteristic requires a large filter inductor and a semi-complex regulator, because of the difficulty of controlling the motor solely by current.

The *pulse-width-modulated* (PWM) inverter drive uses a diode rectifier to provide a constant d.c. voltage. The inverter section, therefore, controls both voltage and frequency. This is done by varying the width of the output pulses, as well as the frequency, in such a way that the effective voltage is approximately sinusoidal.

Because a PWM controller presents a closer simulation of sine-wave power to the motor, fewer power-filter

The unmoving switch

Solid-state switching components are the reason that a.c. and d.c. drives have assumed the lion's share of the adjustable-speed market. The reliability and efficiency of these components are essential to the power conversion processes that they perform in these drives. And as solid-state devices have improved, so have electrical drives. Today's components are vastly better than those on which the solid-state revolution was founded twenty years ago.

Here is a guide to the basic components and how they function:

Transistors—These were the earliest solid-state switching and amplification devices. They have also been the simplest to apply. Transistors are three-terminal devices (base, collector and emitter) in which the base terminal controls the impedance between the other two. The transistor conducts in the forward direction when the base current is high enough, and is turned off when the base current is too low (see sketch).

Until last year, transistors were limited primarily to the smaller a.c. drives, because they did not have the capacity to handle the current loads in drives above 5 hp. Now, however, with the advent of 460-V transistors, the era of high-power transistorized a.c. drive is at hand.

Thyristors—A family of solid-state components turned on by externally applied voltage or current, thyristors were initially used in d.c. drives and high-horsepower a.c. drives, but are beginning to be replaced in a.c. systems by transistors. The limitation of thyristors is that they must be turned off—a necessity in the complex switching required in a.c. controllers.

Silicon-controlled rectifier—The SCR is a type of thyristor that has three terminals: an anode, a cathode and a gate. It is normally off (nonconducting) until a small "trigger" voltage is applied at the gate terminal (see sketch), then it begins to conduct in the forward direction. The problem is, once switched on, the SCR cannot be switched off by a negative signal at the gate; it can only be switched off by interrupting the anode current. In a d.c. drive, this occurs automatically when the a.c. line current goes from positive to negative. However, in an a.c. drive, "forced commutation" is necessary. At the present time, SCRs are used in d.c. drives and in the medium-to-high horsepower (above 20 hp) a.c. drives.

Gate-turnoff thyristor—The GTO is similar to the SCR except that it can be turned off by a negative signal on the gate terminal. GTOs are now used in d.c. and a.c. drives under 20 hp.

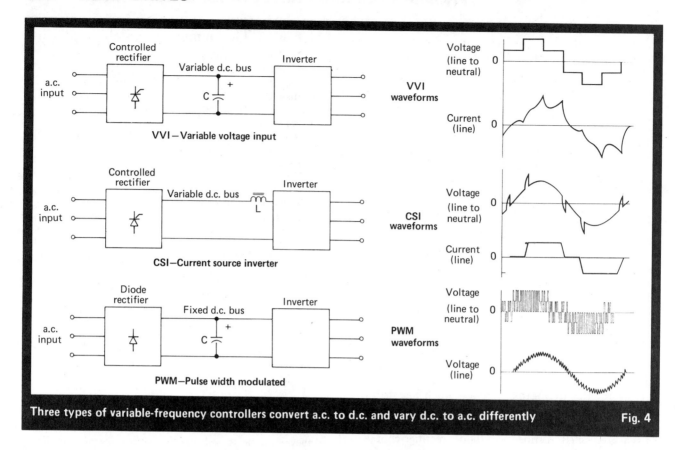

Three types of variable-frequency controllers convert a.c. to d.c. and vary d.c. to a.c. differently Fig. 4

components are required. However, the complex switching waveforms in the inverter require the most complex regulator of the drive types being discussed, and losses due to switching can be high.

Each type of drive has specific advantages:

- At full speed and full load, at which point the drive efficiency is most critical due to the large amount of power handled by the drive, the three foregoing types of adjustable-frequency drives are reasonably close in efficiency—typically 85-90%, including both the controller and motor.

- Efficiencies of all three types of drives vary, depending on horsepower rating and operating conditions. Higher-horsepower drives tend to have higher efficiencies, along with those operating closest to their maximum design rating.

- Motor losses are a function of load current, which is the same regardless of drive type.

- The CSI controller tends to maintain higher efficiency than the others as operating speed is reduced. Commutation losses (those associated with "commutating," or shutting off, the thyristors in the inverter section, a major contributor to total controller losses) tend to vary proportionally to torque and current.

For precise speed regulation

Combining a solid-state controller with a synchronous reluctance motor or a permanent-magnet-rotor motor provides controlled speed on the order of less than 0.5% deviation from set speed. Controller regulator modification can decrease this to less than 0.05% for precise metering in critical control operations.

Beyond base speed

The speed of an a.c. motor is proportional to the signal frequency supplied by the controller. For most applications, this ranges from 6 Hz (at starting) up 60 Hz (base speed). Although a.c. drives can be operated above base speed, they can only deliver *constant horsepower* in this range, and torque decreases as speed increases (Fig. 5).

Frequency signal from controller governs a.c. motor speed Fig. 5

PROPER CHOICE OF ADJUSTABLE-SPEED DRIVES

More than one motor

A distinct advantage of a.c. drives is in applications that require more than one motor to be driven at the same speed or at speeds proportional to each other. Examples include conveyor drives powered by more than one motor, multiple conveyors feeding one another, and container-filling machines that fill and simultaneously move containers. A single a.c. drive, provided that it can accommodate the maximum current demanded by multiple motors, will drive all the motors at the same speed, with each motor sharing the load equally.

When the speeds of two motors must be held at a precise ratio, a.c. drives are again the choice. An example is a blending system combining two materials at a fixed ratio. This is achieved by driving the two motors with separate controllers whose frequency outputs are regulated by a master speed reference. The master reference enables a single operator to keep as many as twelve motors operating at speeds proportional to each other.

After the fact—retrofitting

A special advantage of a.c. controllers is the ease by which they can be retrofitted to existing motors to convert them into adjustable-speed systems. This enables the motor speed to be reduced when maximum output is not required, thus saving energy.

A suitably sized controller replaces the constant-speed drive, and is mated to a transducer that senses some variable of the flow stream, such as pressure or temperature. The inverter controls motor speed in accordance with demands from the transducer.

Retrofitting is particularly beneficial in geographic areas where power costs are high and the equipment involved runs constantly. In a recent retrofitting of ventilators in a large pump system, a constant-speed system was converted into an adjustable-speed one. The initial cost for two controllers, associated transducers, and labor for conversion was $130,000. However, this cost was projected to be recovered in less than one year, based on energy savings at a rate of 6¢/kWh.

In another application, a centrifugal pump that was driven by a 100-hp induction motor and speed regulated with an eddy current clutch was retrofitted with a solid-state drive system. Operating at 8,000 h/yr, at an energy cost of 7¢/kWh, the system had a total energy cost of $25,984. However, when the more efficient solid-state system was installed, energy cost fell to $18,424/yr.

Solid-state d.c. drives

There are several reasons for the prevalence of d.c. drives. They are available in an extremely wide range of power ratings, from fractional to several thousand horsepower. They control speed precisely over the entire speed range, from start to maximum speed.

Electronic controllers for d.c. motors are relatively simple, with a large number of options available for specialized applications, such as dynamic and regenerative braking, acceleration and deceleration control, precise torque control, jog, tension control and quick reversing. The drive package (motor and controller), often costs a little more than other drive packages (motor plus mechanical, electromechanical or fluid transmission), but the extreme versatility of d.c. drives makes them highly utilitarian.

Unlike mechanical, electromechanical or fluid drives, d.c. drives vary output speed by altering the shaft speed of the prime mover. Other drive concepts that are basi-

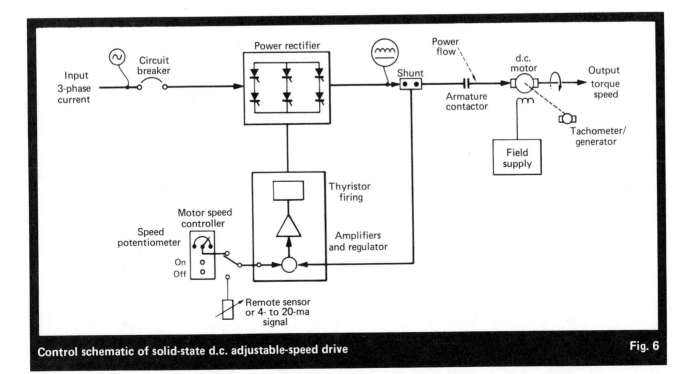

Control schematic of solid-state d.c. adjustable-speed drive   Fig. 6

PUMP DRIVES

A 3-hp motor coupled to a modular d.c. drive control pulls wire through a galvanizing bath Fig. 7

cally controlled couplings between the prime mover and the load do nothing to change the shaft speed of the prime mover, which is usually an a.c. induction motor that operates inherently at only one (synchronous) speed. In such cases, speed is reduced by converting energy into waste heat. But solid-state control drives consume only the energy required to satisfy the demand, plus losses. Thus, when energy conservation is important, d.c. drives are very cost-effective.

Another advantage of these drives is that they can quickly disconnect driving torque from a load, which is important when it is necessary to terminate an operation quickly.

Basic components and control techniques

Earlier, d.c. adjustable-speed drives were ponderous devices that included the motor prime mover, plus a motor-generator set that produced the necessary adjustable voltage. Twenty years ago, however, the advent of the SCR, or thyristor, changed all that. The controller portion of a d.c. drive is now composed of a power unit (rectifier), regulator (signal amplifier), and reference section (operator controls) (Fig. 6).

Using as many as six thyristors, the power unit converts a.c. power to d.c. power that is acceptable to the motor. Controllers for small (to 5 hp) motors often use single-phase full-wave rectifiers; those for larger motors employ three-phase, full-wave systems.

The regulator controls the "firing" of the thyristors in the power unit and, thus, its output.

As with the a.c. drive, the reference section consists of a potentiometer and on/off switch. The d.c. motor itself is made up of two basic elements: an armature-commutator and field. The armature is a wound device that rotates to provide mechanical power. The windings terminate in the copper commutator bars. Power is applied to the armature via carbon brushes, which rub against the commutator. The field is also a wound structure, but is mounted on the inside of the motor frame and provides electromagnetic flux in the thin air gap between the field and armature.

The basic method for changing the speed of the motor is to vary the voltage applied to the armature. This is known as armature, or voltage, control. Within limits, as armature voltage increases, the speed will also increase.

Limitations in the process industries

These drives do, however, have some drawbacks, particularly in many of the chemical process industries.

First, d.c. motors are difficult to make explosion-proof; this can be done, but it takes considerable shielding on the motor and complex ducting of cooling air or an expensive special enclosure. The problem lies at the brush-to-commutator interface, where the d.c. current enters the armature. Because carbon brushes rub across copper commutators, making and breaking contact, sparking occurs. This is the area that must be sealed from volatile chemicals.

Second, open d.c. motors are sensitive to corrosive and particle-laden atmospheres. Corrosives, such as halogens and sulfides, attack the commutator surface, pitting it and impeding the flow of current. Particles, such as silica dust or other abrasives, become embedded in the soft copper brushes and score the commutators.

The brush/commutator area also is the source of most maintenance problems. As current passes between the brushes and armature, a hard oxide film builds up on the commutator. This film is both desirable and undesirable. In place, it minimizes wear on the soft copper commutator. But frequent starts, overloads and corrosive attacks can cause the film to flake, and the flakes to dislodge and become embedded in the carbon brushes. At this point, the oxide film becomes an abrasive and scores the commutator. The depth of the scoring can vary from light "threading" to deep grooving; to remove any such scoring, the commutator must be machined.

Brush monitoring systems are available for many d.c. motors. These take much of the guesswork out of brush maintenance by generating a signal when only 15% of brush life remains.

Close speed regulation

Adjustable-speed d.c. drives can be made to hold speed very close to a set value. For a given voltage (voltage being proportional to speed) to the armature, and with full field applied, the motor will maintain about 95% of the set speed (5% droop), if the load on the shaft varies from about 5 to 100%. This droop can be narrowed by feeding back voltage from the armature. When this feedback is connected to the regulator, speed regulation can be held to about 99% of the set level (1% droop).

For very precise speed regulation, tachometer-generators are used. These, mounted to the motor shaft, generate a voltage that is proportional to speed. When this signal is applied to the regulator, speed droop can be held to 0.1%. Tachometer feedback also enables d.c. drives to operate smoothly at 1/100 of rated speed.

One of the penalties of tachometer feedback is that it

acts as a small drain on motor efficiency. A tachometer is actually a small generator that develops a signal that is proportional to motor speed. Because the tachometer must be driven by the motor, it represents a built-in load (although a small one).

Most d.c. drives can be run at very low speeds, with excellent fidelity to the set speed. However, most of these drives are cooled by internal fans, which are shaft mounted. If high-torque (i.e., high-current) operation is required at low speeds (below about 60% of rated speed), an auxiliary fan may be required to keep the motor from overheating.

Exceptional versatility of d.c. drives

As was mentioned, d.c. drives can be made to perform a number of functions that other types of variable-speed drives cannot. This is one reason why they are so widely used.

Unlike many mechanical, electromechanical, fluid or a.c. drives, d.c. drives can be rapidly reversed. Also, when regulated by certain types of controllers, the motor can be converted to a generator, to act as a dynamic (power spent in a resistor) or regenerative (power reintroduced to the line grid) brake for high-inertia loads, such as centrifuges.

These drives can operate over their rated speed range at constant torque. However, in certain instances, they can even be driven faster than rated (base) speed. This is done by reducing field flux. Operation above base speed is done at constant horsepower (with proportionally diminishing torque), but can range as high as 400% of base speed. And d.c. drives can be designed to control not just speed but also torque.

Applications include driving of extruders, drawing machines, coaters, laminators, winders, and other equipment that must be sensitive to tension or viscosity limitations (Fig. 7). In these cases, controllers are designed to regulate armature current, which is proportional to torque. These drives are typically coupled to a tension or pressure transducer that monitors the process parameter being controlled.

Economy and energy

On a first-cost basis, d.c. drives of a given power rating are usually slightly more expensive than an induction motor coupled to a mechanical, fluid or electromechanical drive. This is because the controls required to regulate the speed in the d.c. motor are more sophisticated than for the other drives. But none of these other drives alters the speed of the prime mover. This is the province of the d.c. and, more recently, a.c. adjustable-speed drives. Coupling systems of the other drives vary output speeds, but they do it at the expense of energy by converting mechanical energy to heat.

The d.c. controllers function at an efficiency of about 98%. Motor efficiency ranges from 87 to 90%. Thus, overall efficiency is typically 86%. This compares favorably with other types of drive systems when prolonged operation at reduced speeds is necessary. In many instances, such as for dryers, extruders and pumps, d.c. drives are actually cheaper in the long run than mechanical, electromechanical or fluid types.

Mechanical drives

Mechanical variable-speed drives are the simplest, least expensive and oldest devices for varying the speed between a driving shaft and a driven shaft. They tend to be lightweight, efficient and easy to maintain. Most function by converting speed into torque, i.e., when speed is reduced, torque is increased. Some of these drives can increase the speed of the output shaft (via sheaves, gears, etc.) to higher than that of the prime mover, but only with a reduction in torque.

The efficiencies of the various mechanical drives depend generally on the amount of internal enthalpic losses, such as friction, and not on component-to-component slippage.

The chief advantages of mechanical drives are simplicity, ease of maintenance and low cost; their chief disadvantages are a moderate degree of maintenance and an inability to disengage load quickly.

Four basic types have evolved and are in use today: belt, chain, wooden block and traction types. The first three are similar in that a continuous web (of rubber-fabric composition, chain, or wooden block belts) transmits torque from one adjustable sheave to another. The last type, the traction drive, is a relatively recent development and transmits torque between a variety of cones, disks, or balls that are pressed tightly together.

Belt drives—light, sturdy, easily serviced

Belt drives are based on a pair of adjustable sheaves and a rubber-fabric composite belt that moves between them. The sheaves can be opened or closed axially, thus changing the "effective pitch" at which the belt contacts the sheaves (Fig. 8). The drive ratio depends on the degree to which one sheave is opened compared to the other. In some cases, only the drive sheave

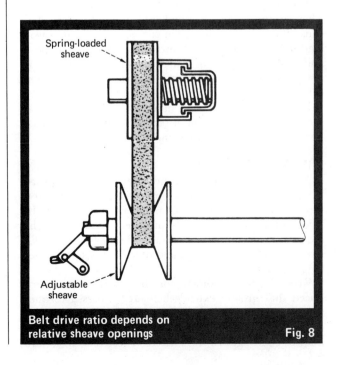

Belt drive ratio depends on relative sheave openings Fig. 8

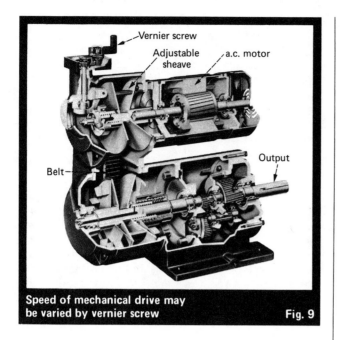

Speed of mechanical drive may be varied by vernier screw — Fig. 9

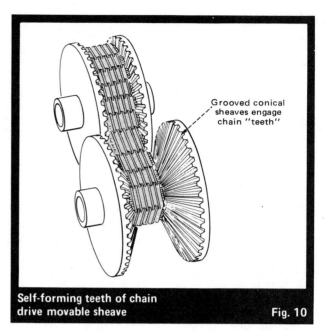

Self-forming teeth of chain drive movable sheave — Fig. 10

is adjustable and the driven sheave has a fixed pitch; in other cases, both are adjustable.

Drive speed may be varied by a vernier-screw mechanism, which is turned (electrically or by hand) to move the sheave halves in or out (Fig. 9). Mechanical and pneumatic shifting arrangements are also available.

Belt drives are available for low to moderate torque applications over a power range to 100 hp (dual-belt systems are usually employed in the higher power ranges). Reduction ratio is frequently as much as 10:1. Maximum speed (without a gear reducer) is typically 4,000 rpm. Many drive packages combine an a.c. induction motor, belt reducer, and gear reducer (fixed ratio), which allows the belt to operate within an optimum speed range and to match a wide range of output speed and power requirements.

The efficiency of belt drives is excellent, usually 95%. They provide good overload and jam protection, because the belt will slip when subject to extreme overloading, which prevents damage to the prime mover, and optimum operating smoothness. Speed control is not precise with these drives; the normal accuracy they can achieve is about 5%.

Belt drives are typically chosen for their light weight, shock tolerance and ease of service, as with conveyors, portable mixers, and other mobile equipment.

Update on belt drives

Although belt drives are the oldest type of adjustable-speed device, their development has not remained static. The critical area of any belt drive is the belt itself. And today, new materials and reinforcement techniques enable single-belt rubber-fabric systems to deliver approximately 50 hp, and dual-belt systems to deliver 100 hp.

Recently, a new design for these drives was introduced that greatly extends belt life. It uses a torque-sensing cam on the variable-speed sheave to provide only the torque necessary to accelerate the load and hold the belt against the sheaves. These belt drives are available in ratings up to 50 hp, and are being sold for fan applications.

Metal-chain drives—for high torque

These drives are similar in principle to belt drives, but use one of two types of metal "belts" instead of rubber-fabric belts. One type of chain has laminated sections in which each link is composed of a number of shims (laminated in the direction of travel) through which slide a stack of hardened steel slats (in the transverse direction). The slats make contact with the movable sheave (Fig. 10). The other type of chain is similar to the conventional sprocket chain, except for having extended pins to contact the sheaves (Fig. 11).

Ratings of chain drives depend to a large extent on the reduction ratio; the greater the ratio, the lower the rating. For high reduction ratios, manufacturers provide derating curves for torque and power.

Chain drives can be more durable (when used with smooth loads), can transmit much higher torques, and can provide better speed accuracy than belt drives. Physically, a chain drive may also be smaller than a comparable belt drive.

However, chain types provide no shock load protection and are only suited to relatively low-speed operation. Excessive slip will destroy the transverse plate edges. Also, they are as much as 50% more expensive.

Applications include driving permanent conveyors, tumbling mills, and other loads that are characterized by high torques, long duty cycles, and low backlash.

Wood-block drives—for rugged service

Wood-block drives, like the belt types, are one of the oldest designs of adjustable-speed drives. This type is also physically similar to belt drives, in that a moving web transmits torque between a drive and driven sheave. Speed is varied by axially adjusting the sheaves to alter the belt contact point. In most cases, both sheaves are adjustable.

Wood-block transmissions have survived in the face

PROPER CHOICE OF ADJUSTABLE-SPEED DRIVES

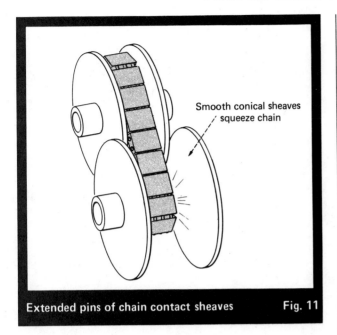

Extended pins of chain contact sheaves — Fig. 11

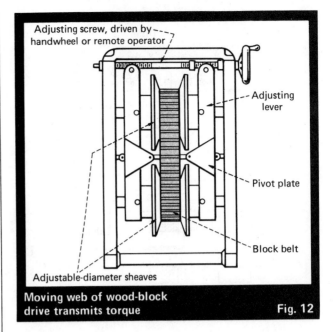

Moving web of wood-block drive transmits torque — Fig. 12

of much newer and faster designs because they are extremely rugged and capable of handling extreme overloads and protecting the prime mover.

The web for these drives is based on a row of transverse rectangular wooden blocks that are tipped with leather. The blocks are bolted to a fabric belt, with spaces between blocks to allow for flexing (Fig. 12). These drives often operate for years under conditions of high backlash, overloads, shock loads and even abrasive environments. They are easily mounted horizontally or vertically. Chief limitations of wood-block drives are physical bulkiness and low input speed (usually below 500 rpm), which usually requires that a speed reducer be placed between the motor and drive.

Applications include grinders and crushers and other devices that are likely to jam or be subject to frequent and heavy shock loads, and that require high starting torques—such as paint mixers.

Electromechanical slip devices

Other established techniques for varying the speed of a motor are the electromechanical clutch and the wound-rotor motor. These employ one of two electromagnetic principles to vary the degree of slip between a drive and the driven element (Fig. 13).

Unlike mechanical drives, such as belt or chain drives, electric clutches do not increase (multiply) torque output as speed is reduced. Also, they are priced on a par with the more expensive mechanical drives.

Precise and fast-changing speed control

Electric clutches have the advantage of rapid change in reduction ratio (some have cycle rates of 1,600/min) from full slip (zero output) to full engagement (almost 100% of prime-mover speed). At full engagement, they draw only enough energy to keep their fields excited, which amounts to less than 1% of the total requirement of most drives.

Another advantage of electric clutches, compared with their mechanical counterparts, is the simple modes by which they can be controlled. The most common mode is, of course, speed control. A tachometer/generator provides d.c. feedback signals from the output shaft to the field exciter and is able to regulate speed to within 0.1% of set value.

In addition, electric clutches can be controlled by a variety of process-related inputs—such as thermistors, pressure transducers, flow transducers and photoelectric cells. In some cases the control signal for these clutches is motor current (which is proportional to torque), so that the output can be torque-regulated.

Finally, the placement of controls for electric clutches is often more convenient than that for mechanical or

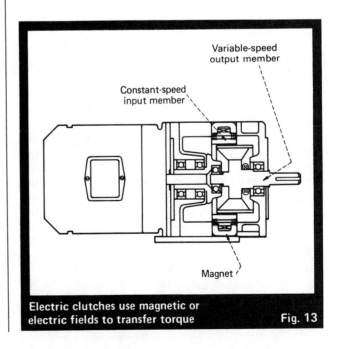

Electric clutches use magnetic or electric fields to transfer torque — Fig. 13

172 PUMP DRIVES

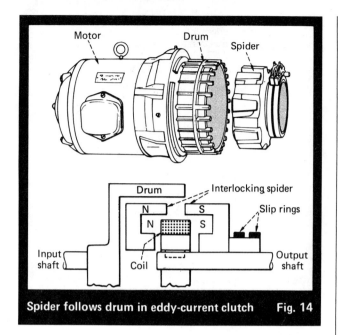

Spider follows drum in eddy-current clutch Fig. 14

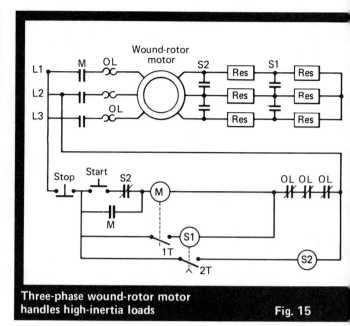

Three-phase wound-rotor motor handles high-inertia loads Fig. 15

fluid drives. Controls are usually located at an operator station, remote from the clutch itself, which is often not the case with mechanical or fluid drives.

One of the drawbacks of electric clutches is that, like fluid drives, when they slip they generate heat; the portion of mechanical energy that does not drive the load is converted to heat.

Most of the smaller clutches—under 50 hp—are air-cooled, and are built with integral cooling fans. Larger clutches—above 100 hp—are usually water-cooled.

Another potential drawback to electric clutches is that they control current for their coils through external slip rings. Although the d.c. signal fed through these slip rings is small, a spark hazard exists.

How eddy-current clutches work

Eddy-current clutches are based on the principle that when a conductor cuts through lines of magnetic flux, currents are induced. When these currents are random, they are called eddy currents. These are undesirable in motors because they increase losses. However, in eddy-current clutches, they are desirable because they generate their own magnetic fields. These fields interact with impressed magnetic fields to produce a force that causes an output member (spider) to follow the driving member (drum) (Fig. 14).

The rotating ferrous drum is driven by a prime mover, usually an a.c. motor. The following output spider is concentric within the drum and carries the coil and slip rings for the d.c. control current. As the drum rotates, the eddy-current field and the main field produce a net flux in the air gap between the drum and spider that is proportional to the coil current. The spider is free to rotate on its own bearings. At full load, slip is 3 to 5%.

The efficiency of an eddy-current clutch is linearly proportional to slip (per cent of input speed). Maximum efficiency—at full speed—is about 96%. However, this falls rapidly when speed is reduced. For this reason, eddy-current clutches tend to be used in applications calling for performance constantly at or near full speed. The greatest speed ratio is usually no more than 2:1.

These clutches are available in a wider variety of sizes and capacities than magnetic-particle clutches. Applications include driving of fans, centrifugal pumps, blowers, and other fluid-propulsion systems that continuously operate at or near maximum speed. They are also found, though not as commonly, in certain constant-torque applications, such as driving extruders and conveyors.

Flexibility from wound-rotor motor

Wound-rotor induction motors are similar to squirrel-cage a.c. motors, except that their rotors are connected to three slip rings. These motors have speed/torque characteristics similar to the conventional induction motor but provide the flexibility of control of starting currents, starting torques and operating speeds (Fig. 15).

Wound-rotor motors vary speed by directing a portion of the energy normally intended for the rotor through external resistors via the slip rings. For most motors of this design, the energy is dissipated as heat by the resistors. This heat represents lost energy and is considered, for design purposes, to be the same as the electromechanical "slip" that takes place in eddy-current clutches.

These motors must be carefully matched to the application if the load is to be slipped continuously, because the resistors build up considerable heat, which must be exhausted to prevent them from being degraded. Wound-rotor motors operate stably at speeds down to about 50% of base speed; thereafter, speed is likely to vary considerably with changes in load.

Applications include driving of pumps for effluent and sludge lines, where the good soft-start characteristics of wound-rotor motors enable them to overcome high inertial loads without overheating. These drives also are applied in certain constant-torque services.

PROPER CHOICE OF ADJUSTABLE-SPEED DRIVES

Fluid adjustable-speed drives

Fluid adjustable-speed drives function like their electromechanical counterparts in that speed reduction is based on the principle of controlled slip between a driving and a driven member. Unlike mechanical belt and chain drives, they do not trade speed reduction for torque multiplication. The degreee of slip corresponds to the degree of speed reduction, and represents lost energy that is dissipated as heat.

Unlike electromechanical drives, however, fluid drives are inherently safe. No metal-to-metal contact takes place—torque is transmitted from input to output through a fluid only—and no sparking can occur.

For large loads

The cost of fluid drives compared with mechanical and electromechanical drives is high; most cannot compete price-wise with these other types of drives in ratings below 25 hp. However, in the 50 to 200-hp range, fluid drives become more competitive. In many cases, their performance characteristics make them desirable for use in the chemical process industries regardless of price.

Fluid drives are characterized by generally high power capacities, smooth torque transmission, large physical packages, and a need to dissipate heat. They are found in applications such as driving of pipeline pumps, crushers, and other equipment that has extremely long duty cycles. Other characteristics include very smooth operation, tolerance for shock loads, ability to withstand periods of stall conditions, inherent safety (they are totally enclosed and have no moving contact), and a tolerance of abrasive atmospheres.

However, fluid drives frequently require considerable "plumbing" for efficient operation. Most need heat exchangers; others, when operated in cold environments, require heaters to keep their working fluid at the proper viscosity (Fig. 16).

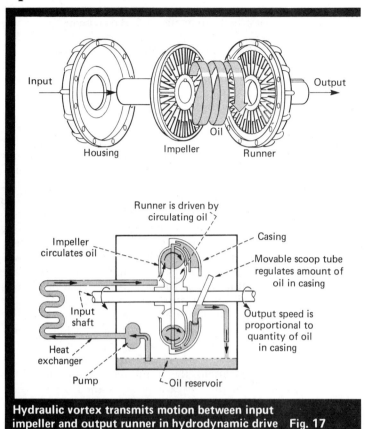

Hydraulic vortex transmits motion between input impeller and output runner in hydrodynamic drive   Fig. 17

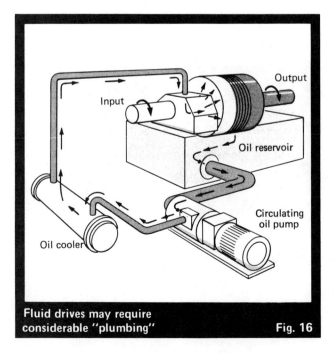

Fluid drives may require considerable "plumbing"   Fig. 16

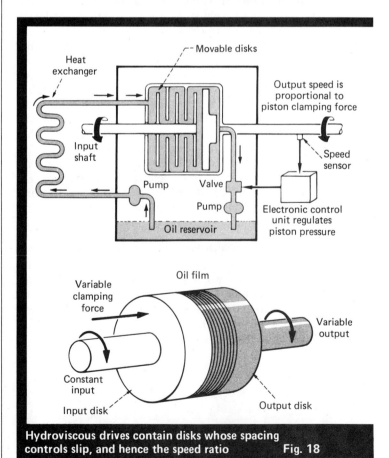

Hydroviscous drives contain disks whose spacing controls slip, and hence the speed ratio   Fig. 18

Two fundamental types of fluid adjustable-speed drives are in common use today in the chemical process industries: hydrodynamic and hydroviscous.

For maximum loads, minimum reduction

Hydrodynamic drives—also known as fluid couplings—are similar to automotive torque converters in that they are based on transmitting motion through a hydraulic vortex between an input impeller and an output runner. Unlike as in torque converters, however, torque is not multiplied. Speed is varied by adjusting the amount of fluid in the toroidal vortex (Fig. 17). As the input impeller turns, it develops the vortex, which in turn produces the pressure on the runner vanes that results in output torque.

When the vortex is small, slip between the input and output is great, and so is the speed reduction. However, this high slip results in considerable heat dissipation and reduced efficiency. Thus, hydrodynamic drives only reach peak efficiency (about 95%) when they are operating at minimum speed reduction, near maximum loads. In this respect, they operate much like eddy-current or magnetic-particle clutches.

These drives should not be operated for long periods at high slip rates. For constant torque conditions, the minimum speed is about 35% of input; for variable torque applications, the minimum is about 20%.

Control of the vortex is accomplished by a "scoop tube," which (as the name implies) removes fluid from the vortex. As the angle of the tube is turned more directly into the fluid stream, it removes more fluid, thus causing more slip. This control mechanism may be connected to an external automatic control that enables the drive to respond to changing load conditions.

The hydrodynamic drive is usually coupled, sometimes directly, to an a.c. motor, which it protects from shock loads, torque overloads and torsional vibration.

One of the advantages of this type of drive is its ability to control load acceleration. Starting torques can be either high or low, depending on the capabilities of the motor. For instance, a drive that must turn a highly loaded centrifuge may be started slowly, so that the inertial torque will not stall the motor. The motor is allowed to achieve full speed before much load is applied.

Applications of hydrodynamic drives include driving of air compressors, ball mills, conveyors, separators and crushers.

For high-horsepower, continuous service

Hydroviscous drives are the choice for extremely-high-horsepower applications requiring continuous or almost continuous operation (in excess of 2,500 h/yr). These drives are not cost-effective compared with mechanical drives in the small power range. Many are built on a custom basis.

Hydroviscous drives are a relative newcomer to the market, first appearing in the early 1960s. Initial models were rated at 200 hp or less and were used to drive medium-sized pumps and fans.

They work as follows: axially spaced disks in a row on an input shaft are interleaved with mating disks attached to an output shaft. The spaces between the disks are filled with oil. When the input shaft turns, a shear force is developed in the film, which results in a following force on the surface of the output disks. This is translated to torque on the output shaft.

Speed control is via pressure applied to the output disks by a piston-powered actuator (Fig. 18). The actuator forces the disks closer together, reducing slip and reducing the speed ratio between the input and output shafts. Actuator pressure is controlled by an external servomechanism.

One important characteristic of hydroviscous drives is fast response time; they can be changed from low load to full load in a fraction of a second. Also, they can operate for years at variable temperature and in highly abrasive atmospheres. A drawback of these drives is poor speed accuracy.

Most common applications for hydroviscous drives are with large pumps, fan drives, and other high-inertia systems that must be operated for years with minimal maintenance.

Acknowledgments

Fig. 8 (variable belt drive, p. 55) and Fig. 14 (eddy-current clutch, p. 58) were obtained from "Controlling Power Transmissions," by Ralph L. Jaeschke, edited by the staff of *Power Transmission Design*, Penton/ITC Publications, Cleveland, Ohio.

Fig. 10 (laminated chain drive, p. 56), Fig. 11 (extended-pin chain drive, p. 57) and Fig. 18 (hydroviscous drive, p. 59) are based on illustrations that appeared in *Machine Design*, Penton/ITC Publications, Cleveland, Ohio.

The author

Thomas R. Doll is Technical Marketing Manager for the A-C V*S Products Group of Reliance Electric Co. (the largest manufacturer of adjustable-speed drives in the world), P.O. Box 608, 55 U.S. Highway No. 46, Pine Brook, NJ 07058. His responsibilities include technical support of the company's field sales force, and other marketing functions. His experience with variable-speed equipment has spanned 15 years. He has attended Stevens Institute of Technology and New York University, and is a member of the American Water Works Assn. and an associate member of the Soc. of Naval Architects and Marine Engineers.

INDEX

Abrasive slurries, seals for, 122-126
Acceleration head, 84
Adjustable-speed drives, 161-174
   electromechanical slip devices, 171-172
   fluid adjustable-speed drives, 173-174
   mechanical drives, 169-171
   solid-state a.c. drives, 163-167
   solid-state d.c. drives, 167-169
Affinity laws, 54-55
Aging effects on centrifugal pumps, 38-40

Bearing and lubrication systems, upgrading, 25-26
Branched-flow service, sizing centrifugal pumps for, 42-44

Cartridge seal, 112
Cast steel for pressure-containing parts, 27-28
Cavitation, low-flow, 6-10
Centrifugal pumps, 23-80
   advantages of standard pumps, 30-31
   choosing plastic pumps, 78-80
   downtime decreased by design improvements, 23-29
   effects of aging on, 38-40
   head-vs.-capacity characteristics of, 32-34
   importance of pump bypasses, 45-49
   low-flow cavitation in, 6-10
   multistage centrifugal pumps, 35-37
   sealless, 118-121
   startup in flashing or cryogenic liquid service, 76-77
   system hydraulics and, 50-72
   systems design, 41-44
   unusual problems with, 73-75
   vertical inline, 4
Computerized methods for pump-related works, 4-6
Controlled-volume pump, 85
Cryogenic liquid service, startup of centrifugal pumps in, 76-77

Designing centrifugal pump systems, 41-44
   design approach, 41-42
   operating cases, 41
   sizing pumps for branched-flow systems, 42-44
Diaphragm pumps, 85, 91-93, 120
Direct-acting pumps, 85
Disk pumps, 14
Displacement (of a rotary pump), 85
Dissolved solids, seals for, 124-125
Donut pumps, 36-37
Double-acting pumps, 85
Double mechanical seals, power consumption of, 127-129

Downtime:
   alternative to Gaede's formula for vacuum pump downtime, 102-103
   upgrading centrifugal pumps for improvement of, 23-29
Duplex pumps, 86

Electromechanical slip devices, 171-172
Electronic motor starters, 155
Energy conservation and pumps, 68-72
   effect of oversizing, 69-71
   effect of specific speed, 68-69
   restoring internal clearances, 71-72
   running one pump instead of two, 71
   variable-speed operation, 71
Energy-saving method of pump selection versus traditional method, 4, 7
Entrained gases, 65
Ethylene-chlorotrifluoroethylene (ECTFE), 79-80
   PVDF versus, 80
Excessive shaft deflection, 23-24

Fiberglass-reinforced plastic (FRP) pumps, 18
Fibrous slurries, seals for, 124
Flow control, 18-19
Flowrates from positive-displacement rotary pumps, predicting, 98-101
Flow repeatability, 85
Fluid adjustable-speed drives, 173-174
Fluoroplastics, 79
Friction head, 51-52

Gaede's formula for vacuum pump downtime, alternative to, 102-103
Gate-turnoff thyristor (GTO), 165
Gear pumps, 95
   sealless, 119-120

Head and system-head curves, 50-54
Head-vs.-capacity characteristics of centrifugal pumps, 32-34
High-temperature metal-bellows seal, 112
Hydraulically stable pumping, 47-48

Leakproof pumps, 12
Linearity, 85
Liquid-hydrocarbon flashing and cryogenic service, startup of centrifugal pumps in, 76-77
Lobe pumps, 95
Lost motion, 85
Low-flow cavitation, 6-10

INDEX

Lubrication systems:
   magnetic shaft seals in, 26-27
   upgrading, 25-26

Magnetic shaft seals in lubrication systems, 26-27
Materials for pumps, 12-13
   listed in order of increasing resistance to wear, 39
   plastics, 16, 18, 78-80
Mechanical seals, 107-139
   for abrasive slurries, 122-126
   demand for higher quality in, 114-117
   power consumption of double mechanical seals, 127-129
   selecting a sealless pump, 118-121
   troubleshooting guide for, 130-139
   user's guide to, 107-113
Mechanical variable-speed drives, 169-171
Metal-bellows seal, 112
Microchip control of electric motors, 144
Modified metal-bellows seal, 112
Modified pusher seal, 111-112
Motors:
   checking pump performance from motor data, 159-160
   new designs for chemical-duty motors, 143-146
   specifying, 147-158
      bearings and vibration, 154-156
      climate, latitude and inertia, 154
      dimensional limitations, 157
      efficiency, 156
      electronic motor starters, 155
      insulation, 152-153
      motor enclosures, 150-152
      motor-starting restrictions, 149-150
      noise, 156-157
      power factor, 157
      protective devices, 158
      special requirements for vertical motors, 157-158
      temperature rise and service factor, 153-154
      testing requirements, 158
      torque requirements, 147-149
      voltage and power supply, 149
Motor starters, electronic, 155
Multiplex pumps, 86
Multispring pusher seal, 111
Multistage centrifugal pumps, 35-37

Net positive suction head available (NPSHA), 84
Net positive suction head required (NPSHR), 84
Nonfibrous abrasive slurries, seals for, 123-124
Nonmetallic-bellows seals, 112

Off-design conditions operation, 65-68

Packed-plunger pumps, 85
Parallel-pump system, 33-34
Piston pumps, 85
Pitot tube pump design, 14
Plastic pumps, 18
   reinforced-, 16
   selection of, 78-80
Plunger pumps, 85
Polypropylene, 78-79

Polytetrafluoroethylene (PTFE), 78, 80
Polyvinylidene fluoride (PVDF), 79
   ECTFE versus, 80
Positive-displacement (PD) pumps, 83-103
   advances in, 13-14
   alternative to Gaede's formula for pump downtime, 102-103
   definition of terms, 83-85
   economic evaluation, 97
   predicting flowrates from, 98-101
   reciprocating pumps, 85-93
   rotary pumps, 93-97
Power consumption of double mechanical seals, 127-129
Power pumps, 85
Pump bypasses, 45-49
Pump drives, 143-174
   checking pump performance from motor data, 159-160
   new designs for chemical-duty motors, 143-146
   selecting the proper adjustable-speed drives, 161-174
      electromechanical slip devices, 171-172
      fluid adjustable-speed drives, 173-174
      mechanical drives, 169-171
      solid-state a.c. drives, 163-167
      solid-state d.c. drives, 167-169
   specifying electric motors, 147-158
Pump efficiency, 85
Pumping principles, 3-4
Pump selection, energy-saving versus traditional method of, 4, 7

Rated capacity, 84
Rated differential pressure, 84
Rated speed (of a rotary pump), 85
Rating curves, affinity laws and, 54-55
Reciprocating pumps, 85-93
   acceleration head, 88-89
   diaphragm pumps, 91-92
   inspection and testing, 91
   mechanical considerations, 89-91
   piston and plunger, 85-87
   pressure pulses, 89
   specification, evaluation and selection, 87-88, 92-93
Reference standards, 15
Reinforced-plastic pumps, 16
Rotary pumps, 85, 93-97
   positive-displacement flowrates for, 98-101

Screw pumps, 93-94
Sealing systems, 12
Sealless pumps, 16-18, 118-121
Shaft deflection, excessive, 23-24
Silicon-controlled rectifier, 165
Simplex pumps, 85-86
Single-acting pump, 85
Single-spring pusher seal, 111
Single-pump system, 33-34
Slip, 85
Solid-state a.c. adjustable-speed drives, 163-167
Solid-state d.c. adjustable-speed drives, 167-169
Solid-state switching components, 165
Specifying electric motors, *see* Motors, specifying

INDEX

Speeds, use of pumps at speeds above 60-Hz, 10-12
Startup of centrifugal pumps in flashing or cryogenic liquid service, 76-77
Static head, 51
Steady-state accuracy, 85
Suction conditions, 55-64
System head, 51
System hydraulics and centrifugal pumps, 50-72
    affinity laws and rating curves, 54-55
    energy conservation and pumps, 68-72
    head and system-head curves, 50-54
    operation at off-design conditions, 65-68
    suction conditions, 55-64
    viscosity and entrained gases, 64-65

Technology, 3-19
    basic principles, 3-4
    leakproof pumps, 12
    low-flow cavitation, 6-10
    materials for pumps, 12-13
    newer models aimed at CPI needs, 16-19
    newer pump types, 14-15
    positive-displacement pump advances, 13-14
    pump sealing systems, 12-13
    reference standards, 15
    testing of pumps, 15
    trends in usage, 4-6
    use at speeds above 60-Hz, 10-12
Testing of pumps, 15
Thermally sensitive fluids, seals for, 125-126
Thyristors, 165
Traditional method of pump selection versus energy-saving method, 4, 7

Transistors, 165
Trends in pump usage, 4-6
Troubleshooting mechanical seals, 130-139
    blistering, 137-138
    causes of failure, 131
    erosion, 135-136
    excessive drive-pin wear, 136
    extrusion, 135
    face distortion, 134-135
    fretting corrosion, 132
    heat checking, 136-137
    leaching, 133-134
    learning via failure analyses, 131
    O-rings
        chemical attack on, 132-133
        overheating, 138-139
    overall chemical attack, 131-132
    oxidation and coking, 139
    spalling, 138
    vaporization, 137
Turndown ration, 85

Upgrading centrifugal pumps to improve downtime, 23-29
User's guide to mechanical seals, 107-113

Vacuum pump downtime, Gaede's formula for, alternative to, 102-103
Vane pumps, 95
Vertical inline centrifugal pumps, 4
Vertical motors, 157-158
Vinyls, 78
Viscosity, 64-65
Volumetric efficiency, 85

*Stolen
From the office of
Neel C. Row*